ISBN 978-1-333-46948-1
PIBN 10508332

1 MONTH OF
FREE
READING

at
www.ForgottenBooks.com

By purchasing this book you are eligible for one month membership to ForgottenBooks.com, giving you unlimited access to our entire collection of over 1,000,000 titles via our web site and mobile apps.

To claim your free month visit:
www.forgottenbooks.com/free508332

Geological Story of the Isle of Wight

BY THE
Rev. CECIL HUGHES, B.A.

LIMITED
LONG ACRE, W.C. 2
1922

Photo by J. Allman Brown, Shanklin.   GORE CLIFF—UPPER GREENSAND WITH CHERT BEDS.

The Geological Story of the Isle of Wight

BY THE
Rev. J. CECIL HUGHES, B.A.

With Illustrations of Fossils by
MAUD NEAL

LONDON :

EDWARD STANFORD, LIMITED

12, 13, & 14 LONG ACRE, W.C. 2.

1922

PREFACE

No better district could be chosen to begin the study of Geology than the Isle of Wight. The splendid coast sections all round its shores, the variety of strata within so small an area, the great interest of those strata, the white chalk cliffs and the coloured sands, the abundant and interesting fossils to be found in the rocks, awaken in numbers of those who live in the Island, or visit its shores, a desire to know something of the story written in the rocks. The Isle of Wight is classic ground of Geology. From the early days of the science it has been made famous by the work of great students of Nature, such as Mantell, Buckland, Fitton, Sedgwick, Owen, Edward Forbes, and others, who have carried on the study up to the present day. Many of the strata are known to geologists everywhere as typical ; several bear the names of the Island localities, where they occur ; some—and those not the least interesting—are not found beyond the limits of the Island. Though studied for so many years, there is no exhausting their interest : new discoveries are constantly made, and new questions arise for solution. To those who have become interested in the rocks of the Island, and the fossils they have found in them, and who wish to learn how to read the story they tell, and to know something of that story, this book is addressed. It is intended to be an introduction to the science of Geology, based on the Geology of the Isle of Wight, yet leading on to some glimpse of the history presented to us, when we take a wider outlook still, and try to trace the whole wondrous path of change from the world's beginning to the present day.

I wish to express my warmest thanks to Miss Maud Neal for the beautiful drawings of fossils which illustrate the book, and to Professor Grenville A. J. Cole, F.R.S., for his kindness in reading the manuscript, and for valuable suggestions received from him. I have also to acknowledge my indebtedness to Mr. H. J. Osborne White's new edition of the *Memoir of the Geological Survey of the Isle of Wight*, 1921 ; and to thank Mr. J. Milman Brown, of Shanklin, for the three photographs of Island scenery, showing features of marked geological interest, and Mr. C. E. Gilchrist, Librarian of the Sandown Free Library, for kindly reading the proofs of the book.

J. CECIL HUGHES.

Mar., 1922.

CONTENTS

CONTENTS

ILLUSTRATIONS OF FOSSILS

PLATE IV.—Facing page 61.

EOCENE   ...   ... CARDITA PLARNICOSTA
TURRITELLA IMBRICATARIA
NUMMULITES LÆVIGATUS
(FUSUS) LEIOSTOMA PYRUS

OLIGOÇENE   ...   ... LIMNÆA LONGISCATA
PLANORBIS EUOMPHALUS
CYRENA SEMISTRIATA

DIAGRAMS

PHOTOGRAPHS

GEOLOGICAL MAP OF THE ISLE OF WIGHT

CHAPTER I

THE ROCKS AND THEIR STORY

WALKING along the sea shore, with all its varied interest, many must from time to time have had their attention attracted by the shells to be seen, not lying on the sands, or in the pools, but firmly embedded in the solid rock of the cliffs and of the rock ledges which run out on to the shore, and have, it may be, wondered sometimes how they got there. At almost any point of the coast of the Isle of Wight, in bands of limestone and beds of clay, in cliffs of sandstone or of chalk, we shall have no difficulty in finding numerous shells. But it is not only in the rocks of the sea coast that shells are to be found. In quarries for building stone and in the chalk pits of the downs we see shells in the rock, and may often notice them in the stones of walls and buildings. How did they get there? The sea, we say, must once have been here. It must have flowed over the land at some time. Now let us think. We are going to read a wonderful story, written not in books, but in the rocks. And it will be much more valuable if we learn to read it ourselves, than if we are just told what other people have made out. We know a thing much better if we see the answers to questions for ourselves than if we are told the answers, and take some one else's word for it. And if we learn to ask questions of Nature, and get answers to them, it will be useful in all sorts of ways all through life. Now, look at the shells in the rock of cliff and quarry. How are they there? The sea cannot have just flowed over and left them. The rock could not have been hard, as it is now, when they

got in. Some of the rocks are sandstone, much like the
sand on the sea shore, but they are harder, and their
particles are stuck together. Does sand on a sea shore
ever become hard like rock, so that shells buried in it are
found afterwards in hard rock? Now we are getting the
key to a secret. We are learning the way to read the
story of the rocks. How? In this way. Look around
you. See if anything like this is happening to-day. Then
you will be able to read the story of what happened long,
long ago, of how this world came to be as it is to-day.
We have asked a question about the sandstone. What
about the clays and the limestone? As before, what is
happening to-day? Is limestone being made anywhere
to-day, and are shells being shut up in it? Are shells in
the sea being covered up with clay,—with mud,—and more
shellfish living on the top of that ; and then, are they, too,
being covered up? So that in years to come they will
be found in layers of clay and stone like those we have
been looking at in quarry and sea cliff?

We have asked our questions. Now we must look
around, and see if we can find the answers. After it has
been raining heavily for two or three days go down to
the marshes of the Yar, and stand on one of the bridges
over the stream. We have seen it flowing quite clear on
some days. Now it is yellow or brown with mud. Where
did the mud come from? Go into a ploughed field with
a ditch by the side. Down the ditch the rain water is
pouring from the field away to the stream. It is thick
with mud. Off the ploughed field little trickles of water
are running into the ditch. Each brings earth from the
field with it. Off all the country round the rain is trickling
away, carrying earth into the ditches and on into the
stream, and the stream is carrying it down into the sea.
Now think. After every shower of rain earth is carried
off the land into the sea. And this goes on all the year
round, and year after year. If it goes on long enough—?

Look a long way ahead, a hundred years,—a thousand,—thousands of years. We shall be talking soon of what takes many thousands of years to do. Why, you say, if it goes on long enough, all the land will be carried into the sea. So it will be. So it must be. You see how the world is changing. You will soon see how it has changed already, what wonderful changes there have been. You will see that things have happened in the world which you never guessed till you began to study Geology.

Now, let us go a bit further. What becomes of all the mud the streams and rivers are carrying down into the sea? Look at a stream coming steeply down from the hills. How it rushes along, rolling pebbles against one another, sweeping everything before it, clearing out its channel, polishing the rocks, and carrying all it rubs off down towards the sea. Now look at a river near its mouth in flat lowland country. It flows now much slower; and so it has not power to bear along all the material it swept down from the hills. And so it drops a great deal; it is always silting up its own channel, and in flood time depositing fresh layers of mud on the flat meadow land,—the alluvial flat,—through which it generally flows in the last part of its course. But a good deal of sediment is carried by the river out to sea. The water of the river, moving slower as it enters the sea, has less and less power to sweep along its burden of sand and mud, and it drops it on the sea bottom,—first the bigger coarser particles like the sand, then the mud; farther out, the finer particles of mud drop to the bottom.

During the exploring cruise of the *Challenger*, under the direction of Sir Wyville Thomson, in 1872-6, the most extensive exploration of the depths of the sea that has been made up to the present time, it was found that everything in the nature of gravel or sand was laid down within a very few miles, only the finer muddy sediments being carried as far as 20 to 50 miles from the land, the very finest of all, under most favourable conditions, rarely extending beyond 150, and never exceeding 300 miles from land into the deep ocean. So gradually layer after layer of sand and mud cover the sea bed round our coasts ; and shells of cockles and periwinkles, of crabs and sea urchins, and other sea creatures that have lived on the bottom of the sea are buried in the growing layers of sand and mud. As layer forms on layer, the lower layers are pressed together, and become more and more solid. And so we have got a good way towards seeing the making of clay and sandstone with shells in them, such as we saw in the sea cliffs and the quarries.

But it is not only rain and rivers that are wearing the land away. All round the coasts the sea is doing the same work. We see the waves beating against the shores, washing out the softer material, hollowing caves into the cliffs, eating away by degrees even the hardest rock, leaving for a while at times isolated rocks like the Needles to mark the former extension of the land. Most people see for themselves the work of the sea, but do not notice so much what the rain and the frost, the streams and the rivers are doing. But these are wearing away the ground over the whole country, while the sea is only eating away at the coast line. So the whole of the land is being worn away, and the sand and mud carried out into the sea, and deposited there, the material of new land beneath the waters.

How do these beds rise up again, so that we find them with their sea shells in the quarry ? Well, we look at the sea heaving up and down with the tides, and we think of the land as firm and fixed. And yet the land also is continually heaving up and down—very slowly,—far too slowly for it to be noticed, but none the less surely. The exact causes of this are not yet well understood, because we know but little about the inside of the earth. The deepest mine goes a very little way. We know that parts of the interior are intensely hot. The temperature in a mine becomes hotter, about 1°F. for every 6oft. we go down on the average. We know that there are great quantities of molten rock in places, which, in a volcanic eruption is poured out in sheets of lava over the land. There are great quantities of water turned into steam by the heat, and in an eruption the steam pours out of the crater of the volcano like the clouds of steam out of the funnel of a locomotive. The people who live about a volcano are living, as it were, on the top of the boiler of a steam engine ; and their country is sometimes shaken up and down like the lid of a kettle by the escaping steam. In such a country the land is often changing its level. A few miles from Naples at Pozzuoli, the ancient Puteoli, may be seen columns of what appears to be an ancient market hall, though it goes by the name of the Temple of Serapis. About half-way up the columns are holes bored by boring shellfish, such as we may find on the shore here at low tide. We see from this that since the building was constructed in Roman times the land has sunk, and carried the columns into the sea, and shellfish have bored into them. Then the land has risen, and lifted the columns out of the sea again.

But it is not only in the neighbourhood of volcanoes that the land is moving. Not suddenly and violently, but slowly and gradually great tracts of land rise and sink. Sometimes the land may remain for a long time nearly stationary. The Southern coasts of England seem to stand at much the same level as in the time of the Romans 1,500 or 2,000 years ago. On the other hand there is evidence which seems to show that the coast of Norway has for some time been gradually rising.

It was thought at one time that the interior of the earth was liquid like molten lava, and that the land we see was a comparatively thin crust over this like the crust of a pie. But it is now believed for various mathematical

B

reasons, that the main mass of the earth is rigid as steel. Still underneath the surface rocks there must be a quantity of semi-fluid matter, like molten rock, and on this the solid land sways about, as we see the ice on a pond sway with the pressure of the skaters on it. So the solid land, pressed by internal forces, rises and falls like the elastic ice, sometimes sinking and letting the sea flow over, then rising again, and bringing up the land from beneath the sea.

Again, as the heated interior of the earth gradually cools by the radiation of the earth's heat into space, it will tend to shrink away from the cooler rocks of the crust. This then, sinking in upon the shrinking interior, will be thrown into folds, like the skin on a shrivelled apple. Seeing, as we often do, layers of rock thrown into numerous folds, so as to occupy a horizontal space far less than that in which they were originally laid down, we can hardly resist the conclusion that shrinkage of the cooling interior of the earth has been a chief cause of the greatest movements of the surface, and of the lateral pressure we so often find the strata to have undergone.

As we study geology we shall find plenty to show that the land does rise and fall, that where now is land the sea has been, that land once stretched where now is sea, though there is still much which is not well understood about the causes of its movements. We have seen how many of the rocks are made in the sea,—the sandstones and the clays,—but there are two other kinds of rocks, about which we must say a little. The first are the Igneous rocks, which means rocks made by fire. These rocks have solidified, most frequently in crystalline forms, from a molten mass. Lava, which flows hot and fluid, from a volcano, and cooling becomes a sheet of solid rock, is an igneous rock. Some igneous rocks solidify under ground under great pressure, and become crystalline rocks such as granite. We shall not find these rocks in the Isle of Wight. We should find them in Cornwall, Wales, and Scotland; and, if we could go deep enough, we should find some such rock as granite underneath the other rocks all the world over. The other rocks, such as the sandstones and clays, are called Sedimentary rocks, because they are formed of sediment, material carried by the sea and rivers, and dropped to the bottom. They are also called Stratified rocks, because they are formed of Strata, *i.e.*, beds or layers, as we see in cliff and quarry.

But we have seen another kind of rock,—the limestones. In Sandown Bay towards the Culvers, bands of limestone run through the dark clay cliffs, and broken fragments lie on the shore, looking like pieces of paving stone. Examining these we find that they are made up of shells, one band of small oysters, the others of shells of other kinds. You see how they have been made. There has been an oyster bed, and the shells have been pressed together, and somehow stuck together, so that they have formed a layer of rock. They are stuck together in this way. The atmosphere contains a small quantity of carbonic dioxide, and the soil a larger quantity, the result of vegetable decomposition. Rain water absorbs some of it, and carries it into the rocks, as it soaks into the ground. This gas has the property of combining with carbonate of lime,—the material of which shells and limestone are made. The bicarbonate of lime so formed is soluble in water, which is not the case with the simple carbonate. Water containing carbonic dioxide soaking into a limestone rock or a mass of shells dissolves some of the carbonate of lime, and carries it on with it. When it comes to an open space containing air, some of the carbonic dioxide is given off, leaving the insoluble carbonate of lime again. So by degrees the hollows are filled up, and a solid layer of rock is formed. Even while gathering in the sea the shell-fragments may be cemented by the deposit of carbonate of lime from sea-water containing more of the soluble bicarbonate than it can hold.

These limestones are examples of rocks which are said to be of organic origin, that is to say, they are formed by living things. Organic rocks may be formed by animal or vegetable growth. Rocks of vegetable origin are seen in the coals. A peat bog is composed of a mass of vegetable matter, chiefly bog moss, which for centuries has been growing and accumulating on the spot. At the bottom of the bog will frequently be found trunks of oak, or other trees, the remains of a forest of former days. The wood has undergone chemical changes, has lost much of its moisture, and often become very hard, as in bog oak. Beds of coal have been formed by a similar process, on a much vaster scale, and continued much longer. The remains of ancient forests have been buried under sand stones and other rocks, have undergone chemical change, and been compressed into the hard solid mass we call coal. Fossil wood, which has not reached the stage of hard coal, but forms a soft brown substance, is called lignite. This is of frequent occurrence in various strata in the Isle of Wight.

Of organic rocks of animal origin the most remarkable are the chalk, of which we shall speak later, and the coral-reefs, which are found in the warm waters of tropical seas. Sailing over the South Pacific you will see a line of trees—coco-nut trees chiefly—looking as if they rose up from the sea. Coming nearer you see that they grow on a low island, which rises only a few feet above the water. These islands are often in the form of a ring, and look "like garlands thrown upon the waters." Inside the ring is a lagoon of calm water. Outside the heavy swell of the Southern Ocean thunders on the coral shore. If a sounding line be let down from the outer edge of the reef, it will be found that the wall of coral goes down hundreds of feet like a precipice. On an island in the Southern Sea,

Funafuti, a deep boring has been made 1,114 ft. deep. As far as the boring went all was coral. All this mass of coral is formed by living things,—polyps they are called. They are like tiny sea anemones, only they grow attached to one another, forming a compound animal, like a tree with stem and branches, and little sea anemones for flowers. The whole organism has a sort of shell or skeleton, which is the coral. Blocks are broken off by the waves, and ground to a coral mud, which fills up the interstices of the coral; and as more coral grows above, the lower part of the reef becomes, by pressure and cementing, a solid coral limestone. Once upon a time there were coral islands forming in a sea, where now is England. These old coral reefs form beds of limestone in Devon, Derbyshire, and other parts of England. In the Isle of Wight we have no old coral reefs, but we shall easily find fossil corals in the rocks. They helped to make up the rocks, but there were not enough here to make reefs or islands all of coral.

The great branching corals that form the reefs can only live in warm waters. So we see that when corals were forming reefs where now is England the climate must have been warm like the tropics. That is a story we shall often read as we come to hear more about the rocks. We shall find that the climate has often been quite warm as the tropics are now: and we shall also read another wonderful story of a time when the climate was cold like the Arctic regions.

Chapter II.

THE STRUCTURE OF THE ISLAND.

THE best place to begin the study of the Geology of the Isle
of Wight is in Sandown Bay.   North of Sandown, beyond
the flat of the marshes, are low cliffs of reddish clay, which
has slipped in places, and is much covered by grass.   At
low tide we shall see the coloured clays on the shore, unless
the sand has covered them up.   Variegated marls they
are called—*marl* means a limy clay, *loam* a sandy clay;
and very fine are the colours of these marls, rich reds and
purples and browns.   Beyond the little sea wall below
Yaverland battery we come to a different kind of clay
forming the cliff.   It is in thin layers.   Clay in thin
layers like this is called *shale*.   Some of these shales are
known as paper shales, for the layers are thin almost like
the leaves of a book.   The junction of the shales with the
marls is quite sharp, and we see that the shales rest on the
coloured marls, not horizontally, but sloping down towards
the North.   Bands of limestone and sandstone running
through the shales, and a hard band of brown rock which
runs out on the shore as a reef, slope in the same direction.
As we pass on by the Red Cliff to the White Cliffs we
notice that the strata slope more steeply the further North
we go.   We have seen that these strata were laid down
layer by layer at the bottom of the sea.   If we find a lot
of things lying one on top of another, we may generally
conclude that the ones at the bottom were put there first,
then the next, and so on to the top.   And this will
generally be true with regard to the rocks.   The lowest
rocks must have been laid down first, then the next, and

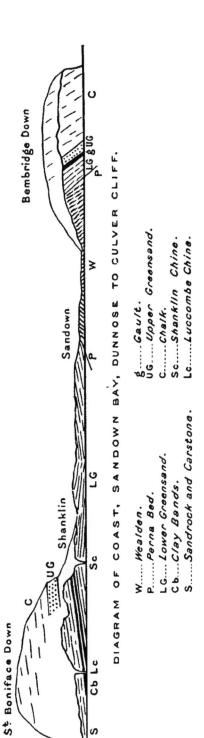

St. Boniface Down   Shanklin   Sandown   Bembridge Down

DIAGRAM OF COAST, SANDOWN BAY, DUNNOSE TO CULVER CLIFF.

W...... Wealden.
P...... Perna Bed.
LG.... Lower Greensand.
Cb.... Clay Bands.
S...... Sandrock and Carstone.

g...... Gault.
UG.... Upper Greensand.
C...... Chalk.
Sc.... Shanklin Chine.
Lc.... Luccombe Chine.

FIG I

so on.   But these layers of shale with shells in them, and
layers of limestone made of shells, must have been laid
down at first fairly flat on the sea floor; but as they
were upheaved out of the sea they have been tilted, so
that we now see them in an inclined position.   And when
we come to the chalk, we should see, if we looked at the
end of the Culver Cliffs from a boat, that the lines of
black flints that run through the chalk are nearly vertical.
The strata there have been tilted up on end.

In describing how strata lie, we call the inclination of
the strata from the horizontal the *dip*.   The direction of
a horizontal line at right angles to that of the dip is called
the *strike.*   If we compare the sloping strata to the roof
of a house, a line down the slope of the roof will mark the
direction of the dip, the ridge of the roof that of the strike.
The strata we are considering dip towards the North;
the line of strike is East and West.

Returning towards Sandown we see the strata dipping
less and less steeply, till near the Granite Fort the rocks
on the shore are horizontal.   Continuing our walk past
Sandown to Shanklin we pass the same succession of rocks
we have been looking at, but in reverse order, and sloping
the other way.   It is not very easy to see this at first, for
so much is covered by building; but beyond Sandown we
see Sandstone Cliffs like the Red Cliff again, the strata
dipping gently now to the south, and in the downs above
Shanklin we see the chalk again.   So we have the same
strata north and south of Sandown, forming a sort of
arch.   But the centre of the arch is missing.   It must
have been cut away.   We saw that the land was all being
eaten away by rain and rivers.   Now we see what they
have done here.   Go up on to the Downs, and look over
the central part of the Island.   We see two ranges of
downs running from east to west,—the Central Downs of
the Island, a long line of chalk down 24 miles from the
Culver Cliff on the east to the Needles on the west; and

the Southern Downs along the South Coast from
Shanklin to Chale. In the Central Downs the chalk
rises nearly vertically, and turns over in the beginning
of an arch towards the South. Then comes a big gap,
and the chalk appears again in the Southern Downs
nearly horizontal, sloping gently to the south. The
chalk was once joined right across the central hollow,
where now we see the villages of Newchurch, Godshill,
and Arreton. All that enormous mass of rock that once
filled the space between the downs has been cut away by
running water.

An arch of strata like this, ∩, such as the one we are
looking at, is called an *anticline*. When the arch is re-
versed, like this **U**, it is called a *syncline*. Looking north
from the Central Downs over the Solent we are looking
at a syncline. The chalk, which dips down at the Culvers
and along the line of the Central Downs, runs like a trough
under the Solent, and rises again, as we see it on the other
side, in the Portsdown Hills.

We might suppose the top of an anticlinal arch would
be the highest part of the country; that, even if rain
and running water have worn the country down, that
would still stand highest, and be worn down least. But
there are reasons why this need not be so. For one thing,
when the horizontal strata are curved over into an arch,
they naturally crack just at the top of the curve, so

and into the cracks the rain gets, and so a stream is
started there, which cuts down and widens its channel,
and so eats the land away. Again, the rising land only
emerges gradually from the sea, and the sea may cut off
the top of the arch before it has risen out of its reach.
Moreover on the higher land the fall of rain and snow is
greater, and the frosts are more severe; so that it is just

there that the forces wearing down the land are most effective.

We must notice another thing which happens when rocks are being upheaved and bent into curves. The strain is very great, and sometimes the strata crack and one side is pushed up more than the other. These cracks are called *faults*. At Little Stairs, about half way between Sandown and Shanklin, two or three faults may be seen in the cliff. The effect of two of the faults may be easily seen by noticing the displacement of a band of rock stained orange by water containing iron. The strata are thrown down towards the north about 8 ft. A third fault, the effect of which is not so evident at first sight, throws the strata down roughly 50 ft. to the south. These are only small faults, but sometimes faults occur, in which the strata have been moved on opposite sides of the fault thousands of feet away from one another. We might think we should see a wall of rock rising up on the surface of the ground where a fault occurs ; but the faults have mostly taken place ages ago ; and, when they do happen, the rocks are generally moved only a little way at a time. Then after a while another push comes on the rocks, and they shift again at the same place, and go a bit further   All this time frost and rain and rivers are working at the surface, and planing it down ; so that the unevenness of the surface caused by faults is smoothed away ; and so even a great fault does not show at the surface.

As we follow the Sandown anticline westward it gradually dies away, the upheaved area being actually a long oval—what we may call a turtle-back. As the Sandown anticline dies out, it is succeeded by another a little further south, the Brook anticline. There are in fact a series of these east and west anticlines in the Island and on the adjacent mainland, caused by the same earth movement. As a consequence of the arching of the strata we find the lowest beds we saw in Sandown Bay running out again on the west of the Island in Brook Bay, and a general correspondence of the strata on the east and west of the Island ; while, as we travel from Sandown or Brook northward to the Solent, we come to continually more recent beds overlying those which appear to the south of them.

When, as in the south side of our central downs, the strata are sharply cut away by denudation, we call this an *escarpment*. The figure shows the structure of the Sandown anticline we have described. We must now examine the rocks more closely, beginning with the lowest strata in the Island, and try to read the story they have to tell.

CHAPTER III

THE WEALDEN STRATA: THE LAND OF THE IGUANODON

THE lowest strata in the Isle of Wight are the coloured marls and blue-grey shales we have already observed in Sandown Bay, which run through the Island to Brook Bay. They are known as the Wealden Strata, because the same strata cover the part of Kent and Sussex called the Weald. They consist of marls and shales with bands of sandstone and limestone. The marls and shales in wet weather become very soft, and flow out on to the shore, causing large slips of land.* Now, what we want to find out is what the world was like ages ago, when these Wealden Strata were being formed. We have learnt something of how clays and sandstones and lime-stones are formed: to learn more we must see what sort of fossils we can find in these rocks. "Fossil" means something dug up; and the word is generally used for remains of animals or plants which we find buried in the rocks. We have seen shells in these strata. These we must examine more closely. And as we walk on the shore we shall find other fossils. In the marls and shales exposed on the shore we are pretty sure to see pieces of wood, black as coal, sometimes quite large logs, often partly covered with shining iron pyrites. Perhaps you say—I hope you do—there must have been land not far away when these marls and shales were forming. Always try

* Blue clays of various geological age, which in wet weather become semi-liquid, and flow out on to the shore, are known in the Island by the local name of *Blue Slipper*.

15

to see what the things we find have to tell us.   The sort
of place where we should be most likely to find wood
floating in the sea to-day would be near the mouth of a
great river like the Mississippi or the Amazon,—rivers
which bring down numerous logs of wood from the forest
country through which they flow.

. Examine the shales and limestone bands.   On the
surface of some of the paper-shales are numbers of small
round or oval white spots.   They are the remains of shells
of a very minute crustacean, Cypris and Cypridea, from
which the shales are known as Cyprid shales.   In other
bands of shale are quantities of a bivalve shell called
Cyrena.   There is a band of limestone made up of Cyrena
shells, containing also little roundish spiral shells called
Paludina.*   This limestone resembles that called Sussex
or Petworth Marble, which is mainly composed of shells
of Paludina, but some layers also contain bivalve shells.
It is hard enough to take a good polish, and may be seen,
like the similar Purbeck marble, in some of our grand old
churches.   Another band of limestone running through
the shales is made up of small oysters (*Ostrea distorta*).

We shall see fossil shells best on the *weathered* surfaces
of rocks, *i.e.*, surfaces which have been exposed to the
weather.   One beginning geological study will probably
think we shall find fossils best by looking at fresh broken
surfaces of rock.   This is not so.   If you want to find
fossils, look at the rock where it has been exposed to the
weather.   The action of the weather—rain, carbonic
dioxide in the rain water, etc.—is to sculpture the surface
of the rock, so that the fossils stand out in relief.   A
weathered surface is often seen covered with fossils, when
a new broken one shows none at all.

Many of the shells in the limestones are very like shells
which are found at the present day.   We must know

* The name now adopted is *Viviparus*.   There is also a band of
ferruginous limestone mainly composed of *Viviparus*.

where they are found now. Well, these Paludinas are a kind of freshwater snail ; and, in fact, all the shells we find in the Wealden strata are freshwater shells, till we come near the top, and find the oysters, which live in salt or brackish water. There were quantities in Brading Harbour in old days, before it was reclaimed from the sea. Now, this is a very important point, that our Wealden shells are freshwater shells. For what does it tell us? Why, we see that the first strata we have come to examine were not laid down in the sea at all. Then where were they formed? They seem to be the Delta of a great river, long since passed away, like the Nile, the Amazon, or the Niger at the present day. When these great rivers near the sea, they spread out in many channels, and deposit the mud they have brought down over a wide area shaped like a V, or like the Greek letter Δ (Delta). Hence we speak of the Delta of the Nile. Some river deltas are of immense size. That of the Niger, for instance, is 170 miles long, and the line where it meets the sea is 300 miles long. Our old Wealden river must have been a great river like the Niger, for the Wealden strata stretch,—often covered up for a long way by later rocks, then appearing again,—as far as Lulworth on the Dorset coast to the west, into Buckinghamshire on the north, while to the north east they not only cover the Weald, but pass under the Straits of Dover into Belgium, and very similar strata are found in Westphalia and Hanover. The ancient river delta must have been 200 miles or more across.

You must not think this great river flowed in the Island of England as it is to-day. England was being made then. This must have been part of a great continent in those days, for such a great river to flow through, and form a delta of such size. We cannot tell quite what was the course of this river. But to the north of where we are now must have stretched a great continent, with

chains of lofty mountains far away, from which the head
waters of the river flowed. Near its mouth the river
broke up into many streams, separated by marsh land ;
while inside the sand banks of the sea shore would be
large lagoons as in the Nile delta at the present day.
In these waters lived the shellfish whose shells we are
finding. And flowing through great forests the river carried
down with it logs of wood and whole trees, and left them
stuck in the mud near its mouths for us to find to-day.

What kind of trees grew in the country the river came
from ?   Well, there were no oaks or beeches, no flowering
chestnuts or apples or mays. But there were great
forests of coniferous trees ; that is trees like our pines
and firs, cedars and yews, and araucarias ; and there were
cycads—a very different kind of tree, but also bearing
cones—which you may see in a greenhouse in botanical
gardens. They have usually a short trunk, sometimes
nearly hemispherical, with leaves like the long leaves of a
date palm. They are sometimes called sago trees, for
the trunk has a large pith, which, like some palms, gives
us sago. Stems of cycads, covered with diamond-shaped
scars, where the leaf stalks have dropped off, are found
in the Wealden deposits. Most of the wood we find is
black and brittle.   Some, however, is hard as stone, where
the actual substance of the wood has been replaced by
silica, preserving beautifully the structure of the wood.
Specially noteworthy are fragments of a tree called
Endogenites (or *Tempskya*) *erosa*, because it was at first
supposed to belong to the endogens,—the class to which
the palm bamboo belong ; it is now considered to be
a tree-fern. Many specimens of this wood are remarkably beautiful, when polished, or in their natural condition. Here, by the way, it may be well to explain how
we name animals and plants scientifically. We have
English names only for the commoner varieties. So we
have to invent names for the greater number of living and

extinct animals and plants. And the best way is found to be this. We give a name, generally formed from the Latin—or the Greek—to a group of animals or plants, which closely resemble one another; the group we call a *genus*. Then for the *species*, the particular kind of animal or plant of the group, we add a second name to the first. Thus, if we are studying the apple and pear group of fruit trees, we call the general name of the group *Pyrus*. Then the crab apple is *Pyrus malus*, the wild pear *P. communis*, and so on. So that when you arrange any of your species, and put down the scientific names, you are really doing a bit of classification as well. You are arranging your specimens with their nearest relations.

To return to our ancient river. With the logs and trunks of trees, which the river brought down, came floating down also the bodies of animals, which had lived in the country the river flowed through. What kind of animals? Very wonderful animals, some of them, not like any living creature that lives to-day. By the time they reached the mouth of the river the bodies had come to pieces, and their bones were scattered about the river mouth. On the shore where we are walking we may find some of these bones. But it is rather a chance whether we find any in any one walk we take. The best time to find them is when rough seas in winter have washed some out of the clay, and left them on the shore. It is only rarely that large bones are found here; but you should be able to find some small ones fairly often. The bones are quite as heavy as stone, for all the pores and cavities have been filled with stone, generally carbonate of lime, in the way we explained in describing the formation of beds of limestone. This makes them quite different from any present-day bones that may happen to lie on the shore. So that you cannot mistake them, if once you have seen them. They are bones of great reptiles,—the class of creatures to which lizards and crocodiles belong.

But these were much larger than crocodiles, and quite peculiar in their appearance. The principal one was the Iguanodon. He stood on his hind legs like a kangaroo, with a great thick tail, which may have helped to support him. When full grown he stood about 14 ft. high. You may find on the shore vertebræ, *i.e.*, joints of the backbone, sometimes large, sometimes quite small if they come from the end of the tail. I have found several here about 5 inches long by 4 or 5 across. A few years ago I found the end of a leg bone almost a foot in diameter. Dr. Mantell, a great geological explorer in the days when these reptiles were first discovered about 80 years ago, estimated from the size of part of a bone found in Sandown Bay that one of these reptiles must have had a leg 9 ft. long. It was a long time after the bones of these creatures were first found before it was known what they really looked like. The animals lived a long way from here, and by the time the river had washed them down to its mouth the skeletons were broken up, and the bones scattered. At last a discovery was made, which told us what the animals were like. In a coal mine at Bernissart in Belgium the miners found the coal seam they were following suddenly come to an end, and they got into a mass of clay. After a while it was seen what had happened. They had struck the buried channel of an old river, which in the Wealden days had flowed through and cut its channel in the coal strata, which are much older still than the Wealden. And in the mud of the ancient buried river what should they come upon but whole skeletons of Iguanodons. In the days of long ago the great beasts had come down to the river to drink, and had got "bogged" in the soft clay. The skeletons were carefully got out, and set up in the Museum at Brussels. Without going so far as that, you may see in the Natural History Museum in London, or the Geological Museum at Oxford, a facsimile of one of these skeletons, large as life, and have

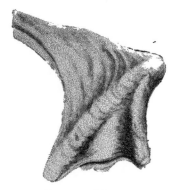

PERNA MULLETI

MEYERIA VECTENSIS
(ATHERFIELD LOBSTER)

PANOPÆA PLICATA

TEREBRATULA SELLA

CYRENA LIMESTONE

IGUANODON VERTEBRA

WEALDEN AND LOWER GREENSAND

some idea of the sort of beast the Iguanodon was. I should tell you why he was so named. Before it was known what he was like in general form, it was found that his teeth, which are of a remarkable character, were similar to those of the Iguana, a little lizard of the West Indies. So he was called Iguanodon,—an animal with teeth like the Iguana (fr. *Iguana*, and Gk. ὀδοῦς g. ὀδόντος a tooth). He was quite a harmless beast, though he was so large. He was a vegetarian. There were other great reptiles, more or less like him, which were also vegetable feeders. But there were also carnivorous reptiles, generally smaller than the herbivorous, whose teeth tell us that they preyed on other animals.

Those were the days of reptiles. Now the earth is the domain of the mammalia. But then great reptiles like the Iguanodon wandered over the land ; great marine reptiles, such as the Plesiosaurus, swam the waters ; and wonderful flying reptiles, the Pterodactyls, flew the air. Some species of these were quite small, the size of a rook : one large species found in the Isle of Wight had a spread of wing of 16 feet. Imagine this strange world,—its forests with pines and monkey puzzles and cycads,—ferns also, of which many fragments are found,—its great reptiles and little reptiles, on land, in the water and the air. Were there no birds? Yes, but they were rare. From remains found in Oolitic strata,—somewhat older than the Wealden,—we know that birds were already in existence ; and they were as strange as anything else. For they had jaws with teeth like the reptiles. They had not yet adopted the beak. And instead of all the tail feathers starting from one point, as in birds of the present day, these ancient birds had long curving tails like reptiles, with a pair of feathers on each joint. Birds of similar but slightly more modern type have been found in Cretaceous strata (to which the Wealden belongs) in America, but so far not in strata of this age in Britain.

Among other objects of interest along this Wealden shore may be noticed a curious transformation which has affected the surface of some of the shell limestones after they were formed, which is known as cone-in-cone structure. It has quite altered the outer layer of the rock, so that all trace of the shells of which it consists is obliterated. Numerous pieces of iron ore from various strata lie on the shore. Through most of English history the Weald of Kent and Sussex was the great iron-working district of England. The ore from the Wealden strata was smelted by the help of charcoal made from the woods that grew there, and gave the district its name ;—for *Weald* means " forest." This industry gradually ceased, as the much larger supplies of iron ore found near the coal in the mines of the North of England came to be worked. Iron pyrites, sulphide of iron in crystalline form, was formerly collected on the Sandown shore, and sent to London for the manufacture of sulphuric acid. This mineral is often found encrusting fossil wood. It also occurs as rounded nodules (mostly derived from the Lower Chalk) with a brown outer coat, and often showing a beautiful radiated metallic structure, when broken. (This form is called marcasite.)

As we walk by the edge of the water, we shall see what pretty stones lie along the beach. When wet with the ripples many look like polished jewels. Some are agates, bright purple and orange in colour, some clear translucent chaldedony. We shall have more to say about these later on. They do not come from the Wealden, but from beds of flint gravel, and are washed along the shore. But there are also jaspers from the Wealden. These are opaque, generally red and yellow. There are also pieces of variegated quartz, and other beautiful pebbles of various mineral composition. These are stones from older rocks, which have been washed down the Wealden rivers, and buried in the Wealden strata, to be washed out again after hundreds of thousands of years, and rolled about on the shore on which we walk to-day.

TRIGONIA CAUDATA

TRIGONIA DÆDALEA

GERVILLIA SUBLANCEOLATA

(AMMONITE)
MORTONICERAS ROSTRATUM

NAUTILUS RADIATUS

LOWER AND UPPER GREENSAND

CHAPTER IV

THE LOWER GREENSAND

FOR ages the Wealden river flowed, and over its vast delta laid down its depth of river mud. The land was gradually sinking ; for continually strata of river mud were laid down over the same area, all shallow-water strata, yet counting hundreds of feet in thickness in all. At last a change came. The land sank more rapidly, and in over the delta the sea water flowed. The sign of coming change is seen in the limestone band made up of small oysters near the top of the Wealden strata. Marine life was beginning to appear.

Above the Wealden shales in Sandown Bay may be seen a band of brown rock. It is in places much covered by slip, but big blocks lie about the shore, and it runs out to sea as a reef before we come to the Red Cliff. The blocks are seen to consist of a hard grey stone, but the weathered surfaces are soft and brown. They are full of fossils, all marine, sea shells and corals. The sea has washed in well over our Wealden delta, and with this bed the next formation, the Lower Greensand, begins. The bed is called the Perna bed, from a large bivalve shell (*Perna mulleti*) frequently to be found in it, though it is difficult to obtain perfect specimens showing the long hinge of the valve, which is a marked feature of the shell. Among other shells are a large round bivalve *Corbis* (*Sphæra*) *corrugata*, a flatter bivalve *Astarte*,—and a smaller oblong shell *Panopæa*,—also a peculiar shell of triangular form, *Trigonia*,—one species *T. caudata* has raised ribs running across it, another *T. dædalea* has bands of raised spots.

A pretty little coral, looking like a collection of little stars, *Holocystis elegans*, one of the Astræidæ, is often very sharply weathered out.

Above the Perna bed lies a mass of blue clay, weathering brown, called the Atherfield clay, because it appears on a great scale at Atherfield on the south west of the Island. It is very like the clay of the Wealden shales, but is not divided into thin layers like shale.

Next we come to the fine mass of red sandstone which forms the vertical wall of Red Cliff. Not many fossils are to be found in these strata. Let us note the beauty of colouring of the Red Cliff—pink and green, rich orange and purple reds. And then let us pass to the other side of the anticline, and walk on the shore to Shanklin. Here we see the red sandstone rocks again, but now dipping to the south. You probably wonder why these red cliffs are called Greensand. But look at the rocks where they run out as ledges on the shore towards Shanklin. Here they are dark green. And this is really their natural colour. They are made of a mixture of sand and clay coloured dark green by a mineral called glauconite. Grains of glauconite can easily be seen in a handful of sand,—better with a magnifying glass. This mineral is a compound of iron, with silica and potash, and at the surface of the rock it is altered chemically, and oxide of iron is formed—the same thing as rust. And that colours all the face of the cliff red. The iron is also largely responsible for our finding so few fossils in these strata. By chemical changes, in which the iron takes part, the material of the shells is destroyed.* Near Little Stairs hollows in the rock may be seen, where large oyster shells have been. In some you may find a broken piece of shell, but the shells have been mostly destroyed. Nearer

* Carbonate of lime has been replaced by carbonate of iron, and the latter converted into peroxide of iron. At Sandown oxidation has gone through the whole cliff.

Shanklin we shall find large oysters, *Exogyra sinuata*, in the rock ledges exposed at low tide. Some are stuck together in masses. Evidently there was an oyster bank here. And here the shells have not been destroyed like those in the cliff.

From black bands in the cliff water full of iron oozes out, staining the cliff red and yellow and orange, and trickling down, stains the flint stones lying on the shore a bright orange. At the foot of the cliff you may sometimes see what looks like a bed of conglomerate, *i.e.*, a bed of rounded pebbles cemented together. This does not belong to the cliff, but is made up of the flint pebbles on the shore, and the sand in which they lie, cemented into a solid mass by the iron in the water which has flowed from the cliff. It is a modern conglomerate, and shows us how old conglomerates were formed, which we often find in the various strata. The cement, however, in these is not always iron oxide. It may be siliceous or of other material. The iron-charged water is called chalybeate ; springs at Shanklin and Niton at one time had some fame for their strengthening powers. The strata we have been examining are known as the Ferruginous sands, *i.e.*, iron sands (Lat. *ferrum*, " iron "). Beyond Shanklin is a fine piece of cliff. Look up at it, but beware of going too close under it. The upper part consists of a fine yellow sand called the Sandrock. At the base of this are two bands of dark clay. These bands become filled with water, and flow out, causing the sandrock which rests on them to break away in large masses, and fall on to the beach.

It is clay bands such as these which are the cause of our Undercliffs in the Isle of Wight. Turn the point, and you see exactly how an undercliff is formed. You see a wide platform at the level of the clay, which has slipped out, and let down the sandrock which rested on it. Beyond Luccombe Chine a large landslip took place in 1910, a great mass of cliff breaking away, and leaving a ravine behind partly filled with fallen pine trees. The whole fallen mass has since sunk lower and nearer to the sea. The broken ground overgrown with trees called the Landslip, as well as the whole extent of the ground from Ventnor and Niton, has been formed in a similar way. But the clay which by its slip has produced these is another clay called the Gault, higher up in the strata. At the top of the high cliff near Luccombe Chine a hard gritty stratum of rock called the Carstone is seen above the Sandrock, and above it lies the Gault clay, which flows over the edge of the cliff.

In the rock ledges and fallen blocks of stone between Shanklin and Luccombe many more fossils may be found than in the lower part of the Ferruginous sands.   Besides bands of oysters, blocks of stone are to be found crowded with a pretty little shell called *Rhynchonella*.   There are others with many *Terebratulæ*, and others with fragments of sea urchins.   The Terebratulæ and Rhynchonellæ belong to a curious group of shells, the Brachiopods, which are placed in a class distinct from the Mollusca proper. They were very common in the very ancient seas of the Cambrian period,—the period of the most ancient fossils yet found,—and some, the Lingulæ, have lived on almost unchanged to the present day.   One of the two valves is larger than the other, and near the smaller end you will see a little round hole.   Out of this hole, when the creature was alive, came a sort of neck, which attached it to the rock, like the barnacles.   There is a very hard ferruginous band, of which nodules may be found along the shore, full of beautifully perfect impressions of fossils, though the fossils themselves are gone.   Casts of a little round bivalve shell, *Thetironia minor*, may easily be got out. The nodules also contain casts of Trigonia, Panopœa, etc. A stratum is sometimes exposed on the shore containing fossils converted into pyrites. A long shell, *Gervillia sublanceolata*, is the most frequent.

All the shells we have found are of sea creatures, and show us that the Greensand was a marine formation. But the strata were formed in shallow water not far from the shore. We have learnt that coarse sediment like sand is not carried by the sea far from the coast. And a good deal of the Greensand is coarser than sand. There are numerous bands of small pebbles. The pebbles are of various kinds ; some are clear transparent quartz, bits of rock-crystal more or less rounded by rolling on the shore of the Greensand period. These go by the name of Isle of Wight diamonds, and are very pretty when polished. Another mark of the nearness of the shore when these beds were laid down is the current bedding, of which a good example may be seen in the cliff at the north of Shanklin parade. It is sometimes called false bedding, for the sloping bands do not mark strata laid down horizontally at the bottom of the sea, but a current has laid down layers in a sloping way,—it may be just over the edge of a sandbank. Again notice how much wood is to be seen in the strata. Land was evidently not far off. All along the shore you may find hard pieces of mineralised wood, the rings of growth often showing clearly. Frequently marine worms have bored into them before they were locked up in the strata ; the holes being generally filled afterwards with stone or pyrites.

The wood is mostly portions of trunks or branches of coniferous trees. We also find stems of cycads. There has been found at Luccombe a very remarkable fruit of a kind of cycad. We said that in the Wealden period none of our flowering plants grew. But these specimens found at Luccombe show that cycads at that time were developing into flowering plants. Wonderful specimens of what may almost be called cycad flowers have been found in strata of about this age in Wyoming in America ; and this Luccombe cycad,—called Benettites Gibsonianus, —shows what these were like in fruit. Remains of various cycadeous plants have been found in the corresponding strata at Atherfield ; and possibly by further research fresh knowledge may be gained of an intensely interesting story,—the history of the development of flowering plants.

On the whole the vegetation of the period was much the same as in the Wealden.   But these flowering cycads must have formed a marked addition to the landscape,— if indeed they did not already exist in the Wealden times. The cones of present day cycads are very splendidly coloured,—orange and crimson,—and it can hardly be doubted that the cycad flowers were of brilliant hues.

The land animals were still like the Wealden reptiles. Bones of large reptiles may at times be found on the shore at Shanklin.   Several have been picked up recently. From the prevalence of cycads we may conclude that the climate of the Wealden and Lower Greensand was subtropical.   The existing Cycadaceæ are plants of South Eastern Asia, and Australia, the Cape, and Central America.   The forest of trees allied to pines and firs and cedars probably occupied the higher land.   Turtles and the corals point to warm waters.   The existing species of Trigonia are Australian shells.   This beautiful shell is found plentifully in Sydney harbour.   It possesses a peculiar interest, as the genus was supposed to be extinct, and was originally described from the fossil forms, and was afterwards found to be still living in Australia.

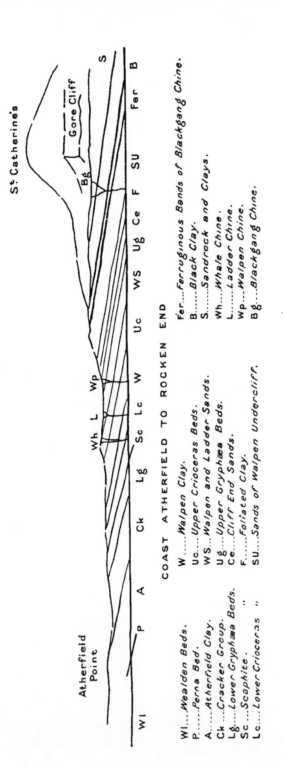

COAST ATHERFIELD TO ROCKEN END

Wl...Wealden Beds.
P......Perna Bed.
A......Atherfield Clay.
Ck...Cracker Group.
Lg...Lower Gryphæa Beds.
Sc...Scaphite   "
Lc...Lower Crioceras "

W......Walpen Clay.
Uc...Upper Crioceras Beds.
WS...Walpen and Ladder Sands.
Ug...Upper Gryphæa Beds.
Ce...Cliff End Sands.
F......Foliated Clay.
SU...Sands of Walpen Undercliff.

Fer...Ferruginous Bands of Blackgang Chine.
B......Black Clay.
S......Sandrock and Clays.
Wh...Whale Chine.
L......Ladder Chine.
Wp...Walpen Chine.
Bg...Blackgang Chine.

FIG. 2

CHAPTER V

BROOK AND ATHERFIELD

To most Sandown Bay is by far the most accessible place in the Island to study the earlier strata ; and for our first geological studies it has the advantage of showing a succession of strata so tilted that we can pass over one formation after another in the course of a short walk. But when we have learnt the nature of geological research, and how to read the record of the rocks, and examined the Wealden and Greensand strata in Sandown Bay, we shall do well, if possible, to make expeditions to Brook and Atherfield, to see the splendid succession of Wealden and Greensand strata shown in the cliffs of the south-west of the Island. It is a lonely stretch of coast, wild and storm-swept in winter. But this part of the Island is full of interest and charm to the lover of Nature and of the old-world villages and the old churches and manor houses which fit so well into their natural surroundings. The villages in general lie back under the shelter of the downs some distance from the shore ; a coastguard station, a lonely farm house, or some fishermen's houses as at Brook, forming the only habitations of man we come to along many miles of shore. Brook Point is a spot of great interest to the geologist. Here we come upon Wealden strata somewhat older than any in Sandown Bay. The shore at the Point at low tide is seen to be strewn with the trunks of fossil trees. They are of good size, some 20 ft. in length, and from one to three feet in diameter. They are known as the Pine Raft, and evidently form a mass of timber floated down an ancient river, and stranded near the mouth, just as happens with great accumulations of timber which float down the Mississippi at the present day. The greater part of the wood has been replaced by stone, the bark remaining as a carbonaceous substance like coal, which, however, is quickly destroyed when exposed to the action of the waves. The fossil trees are mostly covered with seaweed. On the trunks may sometimes be found black shining scales of a fossil fish, *Lepidotus Mantelli*. (A stratum full of the scales of *Lepidotus* has been recently exposed in the Wealden of Sandown Bay.) The strata with the Pine Raft form the lowest visible part of the anticline. From Brook Point the Wealden strata dip in each direction, east and west. As the coast does not cut nearly so straight across the strata as in Sandown Bay, we see a much longer section of the beds. On either side of the Point are coloured marls, followed by blue shales, as at Sandown. To the westward, however, after the shales we suddenly come to variegated marls again, followed by a second set of shales. There was long a question whether this repetition is due to a fault, or whether local conditions have caused a variation in the type of the beds. The conclusion of the Geological Survey Memoir, 1889, rather favoured the latter view, on the ground of the great change which has taken place in the character of the beds in so short a distance, assuming them to be the same strata repeated. The conjecture of the existence of a fault has, however, been confirmed ; for during the last years a most interesting section has been visible at the junction of the shales and marls, where a fault was suspected. The shales in the cliff and on the shore are contorted into the form of a **Z**. The section appears to have become visible about 1904 (it was in the spring of that year that I first saw it), and was described by Mr. R. W. Hooley, F.G.S. (*Proc. Geol. Ass.*, vol. xix., 1906, pp. 264, 265). It has remained visible since.

The Wealden of Brook and the neighbouring coast is celebrated for the number of bones of great reptiles found here, from the early days of geological research, the '20's and '30's of last century, when admirable early geologists, such as Dr. Buckland and Dr. Mantell, were discovering the wonders of that ancient world, to the present time. Various reptiles have been found besides the Iguanodon— the Megalosaurus, a great reptile somewhat similar, but of lighter build, with sabre-shaped teeth, with serrated edges : the Hylæosaurus, a smaller creature with an armour of plates on the back, and a row of angular spines along the middle of the back ; the huge *Hoplosaurus hulkei*, probably 70 or 80 feet in length ; the marine Plesiosaurus and Ichthyosaurus, and several more ; also bones of a freshwater turtle and four types of crocodiles. In various beds a large freshwater shell, *Unio valdensis*, occurs, and in the cliffs of Brook have been found many cones of Cycadean plants. In bands of white sandy clay are fragments of ferns, *Lonchopteris Mantelli*. In the shales are bands of limestone with Cyrena, Paludina, and small oysters, and paper shales with cyprids, as at Sandown. The shore near Atherfield Point is covered with fallen blocks of the limestones.

The Lower Greensand is seen in Compton Bay on the northern side of the Brook anticline. Here is a great slip of Atherfield clay. The beds above the clay are much thinner than at Atherfield, and fossils are comparatively scarce. On the south of the anticline the Perna bed slopes down to the sea about 150 yards east of Atherfield Point, and runs out to sea as a reef. Large blocks lie on the shore, where numerous fossils may be found on the weathered surfaces. The ledges which here run out to sea form a dangerous reef, on which many vessels have struck. There is now a bell buoy on the reef. On the headland is a coastguard station, and till lately there has been a sloping wooden way from the top of the

cliff to bring the lifeboat down. This was washed away
in the storms of the winter 1912-13.

Above the Perna bed lies a great thickness of Atherfield
clay. Above this lies what is called the Lower Lobster
bed, a brown clay and sand, in which are numerous nodules
containing the small lobster *Meyeria vectensis*,—known as
Atherfield lobsters. Many beautiful specimens have been
obtained.

We next come to a great thickness of the Ferruginous
Sands, some 500 feet. The Lower Greensand of Atherfield
was exhaustively studied in the earlier days of geology
by Dr. Fitton, in the years 1824-47, and the different
strata are still referred to according to his divisions. The
lowest bed is the Crackers group about 60 ft. thick. In
the lower part are two layers of hard calcareous boulder-
shaped concretions, some a few feet long. The lower
abound in fossils, and though hard when falling from the
cliffs are broken up by winter frosts, showing the fossils
they contain beautifully preserved in the softer sandy
cores of the concretions. *Gervillia sublanceolata* is very
frequent, also *Thetironia minor*, the Ammonite *Hoplites
deshayesi*, and many more. Beneath and between the
nodular masses caverns are formed, the resounding of the
waves in which has given the name of the " Crackers."
In the upper part of this group is a second lobster bed.

The most remarkable fossils in the Lower Greensand
are the various genera and species of the ammonites and
their kindred. The Ammonite, through many formations,
was one of the largest, and often most beautiful shells.
There were also quite small species. The number of
species was very great. Now the whole group is extinct.
They most resembled the Pearly Nautilus, which still lives.
In both the shell is spiral, and consists of several chambers,
the animal living in the outer chamber, the rest being
air-chambers enabling it to float. The class Cephalopoda,
which includes the Ammonites, the Nautilus, and also the

Cuttle-fish, is the highest division of the Mollusca. The animals all possess heads with eyes, and tentacles around the mouth. They nearly all possess a shell, either external, as in the Nautilus, or internal, as in the cuttle-fishes, the internal shell of which is often washed ashore after a rough sea. The Cephalopods are divided into two orders. The first includes the Cuttle-fish and the Argonaut or Paper Nautilus. Their tentacles are armed with suckers, and they have highly-developed eyes. They secrete an inky fluid, which forms sepia. The internal shell of extinct species of cuttle-fish, of a cylindrical shape, with a pointed end, is a common fossil in various strata, and is known as a Belemnite (Gr. βέλεμνον, " a dart".) The second order includes the Pearly Nautilus of the present day, and the numerous extinct Nautiloids and Ammonoids. The tentacles of the Pearly Nautilus have no suckers ; and the eyes are of a curiously primitive structure,—what may be called a pin-hole camera, with no lens. The shells of the Nautilus and its allies are of simpler form, while the Ammonites are characterised by the complicated margins of the partition walls or septa, by which the shells are sub-divided. The chambers of the fossil Ammonites have often been filled with crystals of rich colours ; and a polished section showing the chambers is then a most beautiful object.*

Continuing along the shore, we come to the Lower Exogyra group, where *Terebratula sella* is found in great abundance. A reef with *Exogyra sinuata* runs out about 350 yards west of Whale Chine. The group is 33 ft. thick, and is followed by the Scaphites group, 50 ft. The beds contain *Exogyra sinuata*, and a reef with clusters of Serpulæ runs out from the cliff. In the middle of the group are bands of nodules containing *Macroscaphites gigas*. The Lower Crioceras bed (16 ft.) follows, and

* Some fine ammonites may be seen at the Clarendon Hotel, Chale,—one about 5 ft. in circumference.

crosses the bottom of Whale Chine. The Scaphites and Crioceras are Cephalopoda, related to the Ammonites; but in this Lower Cretaceous period a remarkable development took place; many of the shells began to take curious forms, to unwind as it were. Crioceras, a very beautiful shell, has the form of an Ammonite, but the whorls are not in contact; thus making an open spiral like a ram's horn, whence its name (Gk. κέρας, ram, κριός, horn). Ancyloceras begins like Crioceras, but from the last whorl continues for some length in a straight course, then bends back again; Macroscaphites is similar, but the whorls of the spiral part are in contact. In Scaphites, a much smaller shell, the uncoiled part is much shorter, and its outline more rounded. It is named from its resemblance to a boat (Gk. σκαφη).*

The Walpen and Ladder Clays and Sands (about 60 ft.) contain nodules with Exogyra and the Ammonite *Douvilleiceras martini*. The dark-green clays of the lower part form an undercliff, on to which Ladder Chine opens. The Upper Crioceras Group (46 ft.), like the Lower, contains bands of Crioceras? also *Douvilleiceras martini*, Gervillia, Trigonia, etc. It must be stated that there is some uncertainty with regard to the ammonoids found in this neighbourhood, Macroscaphites having been described as Ancyloceras, and also sometimes as Crioceras. The discovery of the true Ancyloceras (*Ancyloceras Matheronianum*) at Atherfield is described (and a figure given) by Dr. Mantell; but what is the characteristic ammonoid of the "Crioceras" beds requires further investigation. The neighbourhood of Whale and Walpen Chines is of great interest. Ammonites may be found in the bottom of Whale Chine fallen out of the rock. Red ferruginous nodules with Ammonites lie on the shore, in the Chines, and on the Undercliff, some of the ammonites more or less converted into crystalline spar.

* See *Guide to Fossil Invertebrata*, Brit. Mus. Nat. Hist.

Hard ledges of the Crioceras beds run into the sea. The shore is usually covered deep with sand and small shingle; but there are times when the sea has washed the ledges clear; and it is then that the shore should be examined.

The Walpen and Ladder Sands (42 ft.); the Upper Exogyra Group (16 ft.); the Cliff End Sand (28 ft.); and the Foliated Clay and Sand (25 ft.), consisting of thin alternations of greenish sand and dark-blue clay, follow. Then the Sands of Walpen Undercliff (about 100 ft.); over which lie the Ferruginous Bands of Blackgang Chine (20 ft.). Over these hard beds the cascade of the Chine falls. Cycads and other vegetable remains are found in this neighbourhood. Throughout the Atherfield Greensand fragments of the fern *Lonchopteris* (*Weichselia*) *Mantelli* are found. 220 ft. of dark clays and soft white or yellow sandrock complete the Lower Greensand. In the upper beds of the Greensand few organic remains occur. A beautiful section of Sandrock with the junction of the Carstone is to be seen inland at Rock above Brightstone. The Sandrock here is brightly coloured like the sands of Alum Bay,—though it belongs to a much older formation,—and shows current bedding very beautifully. The junction of the Sandrock and Carstone is also well seen in the sandpit at Marvel.

We have now come to the end of the Lower Cretaceous, in which are included the Wealden and the Lower Greensand. Judged by the character of the flora and fauna, the two form one period, the main difference being the effect of the recession of the shore line, due to the subsidence which let in the sea over the Wealden delta, so that we have marine strata in place of freshwater deposits. But that the plants and animals of the Wealden age still lived in the not distant continent is shown by the remains borne down from the land. These strata are an example of a phenomenon often met with in geology,—that of a great thickness of deposits all laid down in shallow water.

The Wealden of the Isle of Wight are some 700 feet thick, in Kent a good deal thicker, the Hastings Sands, the lower part of the formation, being below the horizon occurring in the Island : the Lower Greensand is some 800 feet thick. In the ancient rocks of Wales, the Cambrian and Silurian strata, are thousands of feet of deposits, mostly laid down in fairly shallow water. In such cases there has been a long-continued deposition of sediment, while a subsidence of the area in which it was laid down has almost exactly kept pace with the deposit. It is difficult not to conclude that the subsidence has been caused by the weight of the accumulating deposit,—continning until some world-movement of the contracting globe has produced a compensating elevation of the area.

CHAPTER VI

THE GAULT AND UPPER GREENSAND

WE have seen how the continent through which the great Wealden river flowed began to sink below the sea level, and how the waters of the sea flowed over what had been the delta of the river, laying down the beds of sandstone with some mixture of clay which we call the Lower Greensand. The next stratum we come to is a bed of dark blue clay more or less sandy, called the Gault. In the upper beds it becomes more sandy and grey in colour. These are known as the " passage beds, " passing into the Upper Greensand. The thickness of the Gault clay proper varies from some 95 to 103 feet. Compared to the mainland the Gault is of small thickness in the Island, though the dark clay bands in the Sandrock mark the oncoming of similar conditions. The fine sediment forming the clay points to a further sinking of the sea bed. In general, we find very few fossils in the Gault in the Island, though it is very fossiliferous on the mainland at Folkestone. North of Sandown Red Cliff the Gault forms a gully, down which a footpath leads to the shore. It is seen at the west of the Island in Compton Bay, where in the lower part some fossil shells may be found.

The Upper Greensand is not very well named, as the beds only partially consist of sandstone, in great part of quite other materials. Some prefer to call the Lower Greensand Vectian, from Vectis, the old name of the Isle of Wight, and the Upper Greensand Selbornian, a name generally adopted, because it forms a marked feature of the country about Selborne in Hampshire.* But, though the Upper Greensand covers a less area in the Isle of Wight than the Lower, it forms some of the most characteristic scenery of the Island. One of the most striking features of the Island is the Undercliff, the undulating wooded country from Bonchurch to Niton, above the sea cliff, but under a second cliff, a vertical wall which shelters it to the North. This wall of cliff consists of Upper Greensand. In a similar way to the small undercliffs we saw at Luccombe, the Undercliff has been formed by a series of great slips, caused here by the flowing out of the Gault clay, which runs in a nearly horizontal band through the base of all the Southern Downs of the Island, the Upper Greensand lying above it breaking off in masses, and leaving vertical walls of cliff. These walls are seen not only in the Undercliff, but also on the northern side of the downs, where they form the inland cliff overhanging a pretty belt of woodland from Shanklin to Cook's Castle, and again forming Gat Cliff above Appuldurcombe. We have records of great landslips at the two ends of the Undercliff, near Bon-church and at Rocken End, about a century ago. But the greater part of the Undercliff was formed by landslips in very ancient times, before recorded history in this Island began. The outcrop of the Gault is marked by a line of springs on all sides of the Southern Downs. The strata above, Chalk and Upper Greensand, are porous and absorb the rainfall, which permeates through till it reaches the Gault Clay, which throws it out of the hill side in springs, some of which furnish a water supply for the surrounding towns and villages.

Where the Upper Greensand is best developed, above the Undercliff, the passage beds are followed by 30 feet of yellow micaceous sands, with layers of nodules of a bluish-grey siliceous limestone known as Rag. The

* Names proposed by the late A. J. Jukes-Browne.

nodules frequently contain large Ammonites and other fossils. Next follow the Sandstone and Rag beds, about 50 feet of sandstone with alternating layers of rag. The sandstones are grey in colour, weathering buff or reddish-brown, tinged more or less green by grains of glauconite. Near the top of these strata is the Freestone bed, a thick bed of a close-grained sandstone, weathering a yellowish grey, which forms a good building stone. Most of the churches and old manor and farm houses in the southern half of the Island are built of this stone. Then forming the top of the series are 24 feet of chert beds,—bands of a hard flinty rock called chert alternating with siliceous sandstone, the sandstone containing large concretions of rag in the same line of bedding. The chert beds are very hard, and where the strata are horizontal, as above the Undercliff, project like a cornice at the top of the cliff. Perhaps the finest piece of the Upper Greensand is Gore Cliff above Niton lighthouse, a great vertical wall with the cornice of dark chert strata overhanging at the top. The thickness in the Undercliff, including the Passage Beds, is from 130 to 160 ft.

The Upper Greensand may be studied at Compton Bay, and at the Culvers ; and along the shore west of Ventnor the lower cliff by the sea consists largely of masses of fallen Upper Greensand, many of which show the chert strata well. In numerous walls in the south of the Island may be seen stone from the various strata—sandstone, blue limestone or rag, and also the chert.

Let us think what was happening when these beds were being formed. The sandstone is much finer than that of the Lower Greensand ; and we have limestones now,— marine, not freshwater as in the Wealden. Marine limestones are formed by remains of sea creatures living at some depth in clear water. And now we come to a new material, chert. It is not unlike flint, and flint is one of the mineral forms of silica. Chert may be called an

impure or sandy flint.　The bands of chert appear to have
been formed by an infiltration of silica into a sandstone,
forming a dense flinty rock, which, however, has a dull
appearance from the admixture of sand, instead of being
a black semi-transparent substance like flint.　But where
did the silica come from?　In the depths of the sea many
sea creatures have skeletons and shells formed of silica
or flint, instead of carbonate of lime, which is the material
of ordinary shells and of corals.　Many sponges, instead
of the horny skeleton we use in the washing sponge, have
a skeleton formed of a network of needles of silica, often
of beautiful forms.　Some marine animalcules, the
Radiolaria, have skeletons of silica.　And minute plants,
the Diatoms, have coverings of silica, which remain like
a little transparent box, when the tiny plant is dead.
Now, much of the chert is full of needles, or spicules, as
they are called, of sponges, and this points to the source
from which some at least of the silica was derived.　To
form the chert much of the silica has been in some manner
dissolved, and deposited again in the interstices of sandstone strata.　We shall have more to say of this process
when we come to speak of the origin of the flints in the
chalk.　Sponges usually live in clear water of some depth ;
so all shows that the sea was becoming deeper when these
strata were being formed.

Along the shore of the Undercliff, Upper Greensand
fossils may be found nicely weathered out.　Very common
is a small curved bivalve shell,—a kind of small oyster,—
Exogyra conica, as are also serpulæ, the tubes formed by
certain marine worms.　Very pretty pectens (scallop
shells) are found in the sandstone.　Many other shells,
Terebratulæ, Trigonia, Panopæa, etc., occur, and several
species of ammonite and nautilus.*　A frequent fossil is

* Of Ammonites, *Mortoniceras rostratum* and *Hoplites splendens*
may be mentioned : and of Pectens, *Neithea quinquecostata* and
quadricostata, Syncyclonema orbicularis, and *Æquipecten asper.*

a kind of sponge, Siphonia.   It has the form of an oblong bulb, supported by a long stem, with a root-like base. It is often silicified, and when broken shows bundles of tubular channels.

In the chert may often be seen pieces of white or bluish chalcedony, generally in thin plates filling cracks in the chert.   This is a very pure and hard form of silica, beautifully clear and translucent.   Pebbles which the waves have worn in the direction of the plate are very pretty when polished, and go by the name of sand agates. They may sometimes be picked up on the shore near the Culvers.

CHAPTER VII

THE CHALK

As we have traced the world's history written in the rocks we have seen an old continent gradually submerged, a deepening sea flowing over this part of the earth's surface. Now we shall find evidence of the deepening of the sea to something like an ocean depth. We are coming to the great period of the Chalk, the time when the material was made which forms the undulating downs of the south-east of England, and of which the line of white cliffs consists, which with sundry breaks half encircles our shores, from Flamborough Head in Yorkshire, by Dover and the Isle of Wight, to Bere in Devon. Across the Channel white cliffs of chalk face those of England, and the chalk stretches inland into the Continent. Its extent was formerly greater still. Fragments of chalk and flint are preserved in Mull under basalt, an old lava flow, and flints from the chalk are found in more recent deposits (Boulder Clay) on the East of Scotland, pointing to a former great extension northward, which has been nearly all removed by denudation. In the Isle of Wight the chalk cliffs of Freshwater and the Culvers are the grandest features of the Island ; while all the Island is dominated by the long lines of chalk downs running through it from east to west. Now what is the chalk ? And how was it made ? The microscope must tell us. It is found that this great mass of chalk is made up principally of tiny microscopic shells called Foraminifera, whole and in crushed fragments. There are plenty of foraminifera in the seas to-day ; and we need not go far to find similar shells.

On the shore near Shanklin you will often see streaks of
what look like tiny bits of broken shell washed into
depressions in the sand.   These, however, often consist
almost entirely of complete microscopic shells, some of
them of great beauty.   The creature that lives in one
of these shells is only like a drop of formless jelly, and yet
around itself it forms a complex shell of surprising beauty.
The shells are pierced with a number of holes, hence their
name (fr. Lat. *foramen*, a hole, and *ferre*, to bear).
Through these holes the animal puts out a number of
feelers like threads of jelly, and in these entangles particles
of food, and draws them into itself.   Now, do we anywhere
to-day find these tiny shells in such masses as to build
up rocks?   We do.   The sounding apparatus, with which
we measure the depths of the sea, is so constructed as to
bring up a specimen of the sea bottom.   This has been
used in the Atlantic, and it is found that the really deep
sea bottom, too far out for rivers and currents to bring
sand and mud from the land, is covered with a white mud
or ooze.   And the microscope shows this to be made up
of an unnumerable multitude of the tiny shells of foraminifera.   As the little creatures die in the sea, their shells
accumulate on the bottom, and in time will be pressed
into a hard mass like chalk, the whole being cemented
together by carbonate of lime, in the way we explained
in describing the making of limestones.   So we find chalk
still forming at the present day.   But what ages it must
take to form strata of solid rock of such tiny shells !   And
what a vast period of time it must have required to build
up our chalk cliffs and downs, composed in large part of
tiny microscopic shells ! With the foraminifera the
microscope shows in the chalk a multitude of crushed
fragments, largely the prisms which compose bivalve
shells, flakes of shells of Terebratula and Rhynchonella,
and minute fragments of corals and Bryozoa.   Scattered
in the chalk we shall also find larger shells and other

remains of the life of the ancient sea. The base of the cliffs and fallen blocks on the shore are the best places to find fossils. Much of the base of the cliffs is inaccessible except by boat. The lower strata may be examined in Sandown and Compton Bays, and the upper in White- cliff Bay. A watch should always be kept on the tide. The quarries along the downs are not as a rule good for collecting, as the chalk does not become so much sculp- tured by weathering.

The deep sea of the White Chalk did not come suddenly. In the oncoming of the period we find much marl—limy clay. As the sea deepened, little reached the bottom but the shells of foraminifera and other marine organisms. How deep the sea became is uncertain : there is reason to believe that it did not reach a depth such as that of the Atlantic.

It is difficult to draw the line between the Upper Greensand and the Chalk strata. Above the Chert beds is a band a few feet thick known as the Chloritic Marl, which shows a passage from sand to calcareous matter. It is named from the abundance of grains of green colour- ing matter, now recognised as glauconite ; so that it would be better called Glauconitic Marl. It is also re- markable for the phosphatic nodules, and for the numerous casts of Ammonites, Turrilites, and other fossils mostly phosphatized, which it contains. This band is one of the richest strata in the Island for fossils. It differs, however, in different localities both in thickness and composition. It is best seen above the Undercliff, and in fallen masses along the shore from Ventnor to Niton. It is finely exposed on the top of Gore Cliff, where the flat ledges are covered with fossil Ammonites, Turrilites, Pleurotomaria, and other shells. The Ammonite (*Schloen- bachia varians*) is especially common. The sponge (*Stauronema carteri*) is characteristic of the Glauconitic Marl. As the edge of the cliff is a vertical wall, none

(PECTEN)

NEITHEA QUINQUECOSTATA

THETIRONIA
MINOR

(AMMONITE)
MANTELLICERAS MANTELLI

RHYNCHONELLA
PARVIROSTRIS

(SEA URCHINS)

MICRASTER COR-ANGUINUM

ECHINOCORYS SCUTATUS
(Internal cast in flint)

LOWER AND UPPER GREENSAND AND CHALK

should try this locality but those who can be trusted to take proper care on the top of a precipice. When a high wind is blowing the position may be especially dangerous.

The Chloritic Marl is followed by the Chalk Marl, of much greater thickness. This consists of alternations of chalk with bands of Marl, and contains glauconite and also phosphatic nodules in the lower part. Upwards it merges into the Grey Chalk, a more massive rock, coloured grey from admixture of clayey matter. These form the Lower Chalk, the first of the three divisions into which the Chalk is usually divided. Above this come the Middle and Upper, which together form the White Chalk. They are much purer white than the lower division, which is creamy or grey in colour. The Chalk Marl and Grey Chalk are well seen at the Culver Cliff, and run out in ledges on the shore. The lower part of this division is the most fossiliferous, and contains various species of Ammonites, Turrilites, Nautilus, and other Cephalopoda. (Of Ammonites *Schloenbachia varians* is characteristic. Also may be named *S. Coupei, Mantelliceras mantelli, Metacanthoplites rotomagensis, Calycoceras naviculare,* the small Ammonoid Scaphites æqualis ; and of Pectens, *Æquipecten beaveri* and *Syncyclonema orbicularis* may be mentioned). White meandering lines of the sponge *Plocoscyphia labrosa* are conspicuous in the lower beds. The Chalk Marl is well shown at Gore Cliff, sloping upwards from the flat ledges of the Chloritic Marl. It may be studied well, and fossils found, in the cliff on the Ventnor side of Bonchurch Cove,—which has all slipped down from a higher level.

The uppermost strata of the Lower Chalk are known as the Belemnite Marls. They are dark marly bands, in which a Belemnite, *Actinocamax plenus,* is found. The hard bands known as Melbourn Rock and Chalk Rock, which on the mainland mark the top of the Lower and Middle Chalk respectively, are neither of them well marked in the Isle of Wight. In the Middle Chalk *Inoceramus labiatus*, a large bivalve shell, occurs in great profusion ; and in the Upper flinty Chalk are sheets of another species, *I. Cuvieri.* It is hardly ever found perfect, the shells being of a fibrous structure, with the fibres at right angles to the surface, and so very fragile.

There is a striking difference between the Middle and Upper Chalk, which all will observe. It consists in the numerous bands of dark flints which run through the Upper Chalk parallel to the strata. The Lower Chalk is entirely, and the Middle Chalk nearly, devoid of flint. Though the line at which the commencement of the Upper Chalk is taken is rather below the first flint band of the Upper Chalk, and a few flints occur in the highest beds of the Middle Chalk ; yet, speaking generally, the great distinction between the Middle and Upper Chalk, the two divisions of the White Chalk, may be considered to be that of flintless chalk and chalk with flints.

Early in our studies we noticed the great curves into which the upheaved strata have been thrown, and that on the northern side of the anticline the strata are in places vertical. This can be well observed in the Culver Cliffs and Brading Down, where the strata of the Upper Chalk are marked by the lines of black flints. In the large quarry on Brading Down the vertical lines of flint can be clearly seen ; and by walking at low tide at White-cliff Bay round the corner of the cliff, or by observing the cliff from a boat, we may see a beautiful section of the flinty chalk, the lines of black flints sloping at a high angle. The flints in general form round or oval masses, but of irregular shape with many projections, and the masses lie in regular bands parallel to the stratification. The tremendous earth movement which has bent the strata into a great curve has compressed the vertical portion into about half its original thickness, and has made the chalk of our downs extremely hard. It has

Photo by J. Milman Brown, Shanklin.] CULVER CLIFFS—HIGHLY INCLINED CHALK STRATA

also shattered the flints in the chalk into fragments. The
rounded masses retain their form, but when pulled out of
the chalk fall into sharp angular fragments, and we find
they are shattered through and through.

Now, what are flints, and how were they formed?
Flints are a form of silica, a purer form than chert, as the
chalk in which they are embedded was formed in the deep
sea, and so we have no admixture of sand. Flints, as we
find them in the chalk, are generally black translucent
nodules, with a white coating, the result of a chemical
action which has affected the outside after they were
formed. Flint is very hard,—harder than steel. You
cannot scratch it with a knife, though you may leave a
streak of steel on the surface of the flint. This hardness
is a property of other forms of silica, as quartz and chalcedony. The question how the flints were formed is a
difficult one. As to this much still remains obscure.
The sea contains mineral substances in solution. Calcium
sulphate and chloride, and a small amount of calcium
carbonate (carbonate of lime) are in solution in the sea.
From these salts is derived the calcium deposited as
calcium carbonate to form the shells of the Foraminifera
and the larger shells in the Chalk. There is also silica in
small quantity in sea water. From this the skeletons of
radiolaria and diatoms and the spicules of sponges are
formed. Now, many flints contain fossil sponges, and
when broken show a section of the sponge clearly marked.
Especially well can this be seen in flints which have lain
some time in a gravel bed formed of flints worn out of the
chalk by denudation. Hard as a flint seems, it is penetrated
by numerous fine pores. The gravel beds are usually stained
yellow by water containing iron, and this has penetrated by
the pores through the substance of the flints, staining them
brown and orange. Many of the stained flints show beautifully the sponge markings,—a wide central canal with
fine thread-like canals leading into it from all sides.

The Chalk Sea evidently abounded in siliceous organisms, and it cannot be doubted that it is from such organisms that the silica was derived, which has formed the masses of flint. Silica occurs in two forms—in a crystalline form as quartz or rock crystal, and as amorphous, *i.e.*, formless or uncrystalline (also called opaline) silica. The siliceous skeletons of marine organisms are formed of amorphous silica. Flint consists of innumerable fine crystalline grains, closely packed together. Amorphous silica is less stable than crystalline, and is capable of being dissolved in alkaline water, *i.e.*, water containing carbonate of sodium or potassium in solution. If the silica so dissolved be deposited again, it is generally in the crystalline form. It seems probable, therefore, that the amorphous silica of the skeletal parts of marine organisms has been dissolved by alkaline water percolating through the strata, and re-deposited as flint.

As the silica was deposited, chalk was removed. The large irregular masses of flint lying in the Chalk strata have clearly taken the place of chalk which has been removed. Water charged with silica soaking through the strata has deposited silica, and at the same time dissolved out so much carbonate of lime. Bivalve shells, originally carbonate of lime, are often replaced, and filled up by flint, and casts of sea urchins in solid flint are common, and often beautiful fossils. This process of change took place after the foraminiferal ooze had been compacted into chalk strata ; and to some extent at any rate, there has been deposition of silica after the chalk had become hard and solid ; for we find flat sheets, called tabular flint, lying along the strata, or filling cracks cutting through the strata at right angles. But in all probability the re-arrangement of the constituents of the strata took place in the main during the first consolidation, as the strata rose above the sea-level, and the sea-water drained out. A suggestion has been made by R. E. Liesegang, of Dresden, to explain the occurrence of the flints in the bands with clear interspaces between, which are such a marked feature of the Upper Chalk. He has shown how " a solution diffusing outward and encountering something with which it reacts and forms a precipitate, moves on into this medium until a concentration sufficient to cause precipitation of the particular salt occurs. A zone of precipitation is thus formed, through which the first solution penetrates until the conditions are repeated, and a second zone of precipitate is thrown down. Zone after zone may thus arise as diffusion goes on." He suggests that the zones of flint may be similar phenomena, water diffusing through the masses of chalk taking up silica till such concentration is reached that precipitation takes place, the water then percolating further and repeating the process.*

The precipitation of silica and replacement of the chalk occurs irregularly along the zone of precipitation, forming great irregular masses of flint, which enclose the sponges and other marine organisms that lay in the chalk strata. Where a deposit of silica has begun, it will probably have determined the precipitation of more silica, in the manner constantly seen in chemical precipitation ; and it would seem that siliceous organisms as sponges have to some extent served as centres around which silica has been precipitated, for flints are very commonly found, having the evident external form of sponges.

It will be well to say something here of the history of the flints as the chalk which contains them is gradually denuded away. Rain water containing carbonic dioxide has a great effect in eating away all limestone rocks, chalk included. A vast extent of chalk, which formerly covered much of England has thus disappeared. The arch of chalk connecting our two ranges of downs has been cut through, and from the top of the downs themselves a

* See *Common Stones*, by Grenville A. J. Cole, F.R.S. 1921.

great thickness of chalk has been removed.   The chalk in the downs above Ventnor and Bonchurch is nearly horizontal.   It consists of Lower and Middle Chalk ; and probably a small bit of the Upper occurs.   But the top of St. Boniface Down is covered with a great mass of angular flint gravel, which must have come from the Upper Chalk.   The gravel is of considerable thickness, perhaps 20 ft., and on the spurs of the down falls over to a lower level like a table-cloth.   It is worked in many pits for road metal.   This flint gravel represents the insoluble residue which has been left when the Chalk was dissolved away.

On the top of the cliffs between Ventnor and Bonchurch, at a point called Highport, is a stratum of flint gravel carried down from the top of the down.   The shore here is strewn with large flints fallen from the gravel. The substance of many of the flints has undergone a remarkable change.   Instead of black or dull grey flint it has become translucent agate, of splendid orange and purple colours, or has been changed into clear translucent chalcedony.   In the agate the forms of fossil sponges can often be beautifully seen.   The colours are due to iron-charged water percolating into the flint in the gravel bed, but further structural changes have altered the form of the silica ; chalcedony having a structure of close crystalline fibres, revealed by polarized light : when variously stained and coloured, it is usually called agate.   Many of these flints, when cut through and polished, are of great beauty.   The main force of the tides along these shores is from west to east ; and so there is a continual passage of pebbles on the shore in that direction.   The flints in Sandown Bay have in the main travelled round from here ; and towards the Culvers small handy specimens of agates and chalcedonies rounded by the waves may be collected.

Photo by J. Milman Brown, Shanklin.]

SCRATCHELL'S BAY—HIGHLY INCLINED CHALK STRATA

The extensive downs in the centre of the Island are largely overspread with angular flint gravel similarly formed to that of St. Boniface. Of other beds of gravel, which have been washed down to a lower level by rivers or other agency we shall have more to say later.

The Chalk strata in the Isle of Wight are of great thickness. In the Culver Cliff there are some 400 feet of flintless Chalk (Lower and Middle Chalk), and then some 1,000 feet of chalk with flints. There is some variation in the thickness of the strata in different parts of the Island, and the amount of the Upper strata, which has been removed by denudation, varies considerably. The average thickness of the white chalk in the Island is about 1,250 feet* Including the Lower Chalk, the maximum thickness of the Chalk strata is 1,630 ft.

The divisions of the chalk we have so far considered depend on the character of the rock : we must say a word about another way of dividing the strata. It is found that in the chalk, as in other strata, fossils change with every few feet of deposit. We may make a zoological division of the chalk by seeing how the fossils are distributed. The Chalk was first studied from this point of view by the great French geologist, M. Barrois, who divided it into zones, according to the nature of the animal life, the zones being called by the name of some fossil specially characteristic of a particular zone. More recently Dr. A. W. Rowe has made a very careful study of the zones of the White Chalk, and is now our chief authority on the subject. The strata have been grouped into zones as follows :—

*1,472 ft. at the western end of the Island, 1,213 ft. at the eastern. —Dr. Rowe's measurements.

	Zones.	Sub-Zones.

Upper Chalk
- Belemnitella mucronata.
- Actinocamax quadratus.
- Offaster pilula { Offaster pilula. / Echinocorys depressus.
- Marsupites testudinarius { Marsupites. / Uintacrinus.
- Micraster cor-anguinum.
- Micraster cor-testudinarium.
- Holaster planus.

Middle Chalk.
- Terebratulina lata.
- Inoceramus labiatus.

Lower Chalk.
- Holaster subglobosus { Actinocamax plenus (at top).
- Schloenbachia varians { Stauronema carteri (at base).

The method of study according to zoological zones is of great interest. The period of the White Chalk was of long duration, and the physical conditions remained very uniform. So that by studying the succession of life during this period we may learn much about the gradual change of life on the earth, and the evolution of living things.

We have seen that the whole mass of the chalk is made up mainly of the remains of living things,—mostly of the microscopic foraminifera. We have seen that sponges were very plentiful in that ancient sea. Of other fossils we find brachiopods—different species of Terebratula and Rhynchonella—a large bivalve *Inoceramus* sometimes very common; the very beautiful bivalve, *Spondylus spinosus*, belemnites, serpulæ; and different species of sea-urchin are very common. A pretty heart-shaped one, *Micraster cor-anguinum*, marks a zone of the higher chalk, which runs along the top of our northern downs. Other common sea urchins are various species of *Cidaris*, of a form like a turban (Gk. *cidaris*, a Persian head-dress); *Cyphosoma*, another circular form; the oval *Echinocorys scutatus*, which, with varieties of the same and allied species, abounds in the Upper Chalk, and the more conical *Conulus conicus*. The topmost zone, that of *B. Macronata*, would yield a record of exuberant life, were the chalk soft and horizontal. There was a rich development of echinoderms (sea urchins and star fishes), but nothing is perfect, owing to the hardness of the rock (Dr. Rowe). The general difference in the life of the Chalk period is the great development of Ammonites and other Cephalopods in the Lower Chalk, and of sea urchins and other echinoderms in the Upper, while the Middle Chalk is wanting in the one and the other. Shark's teeth tell of the larger inhabitants of the ocean that flowed above the chalky bottom.

Many quarries have been opened on the flanks of the Chalk Downs, of which a large number are now disused. They occur just where they are needed for chalk to lay on the land, the pure chalk on the north of the Downs to break up the heavy Tertiary clays, which largely cover the north of the Island ; the more clayey beds of the Grey Chalk on the south of the downs to stiffen the light loams of the Greensand.*

*Dr, A. W. Rowe.

CHAPTER VIII

THE TERTIARY ERA : THE EOCENE

AGES must have passed while the ocean flowed over this part of the world, and the chalk mud, with its varied remains of living things, gradually accumulated at the bottom. At last a change came. Slowly the sea bed rose, till the chalk, now hardened by pressure, was raised into land above the sea level. As soon as this happened, sea waves and rain and rivers began to cut it down. There is evidence here of a wide gap in the succession of the strata. Higher chalk strata, which probably once existed, have been washed away, while the underlying strata have been planed off to an even surface more or less oblique to the bedding-planes. The highest zone of the chalk in the Island (that of *Belemnitella macronata*) varies greatly in thickness, from 150 ft. at the eastern end of the Island to 475 at the western. The latest investigations give reason to conclude that this is due to gentle synclines and anticlines, which have been planed smooth by the erosion which preceded the deposition of the next strata,—the Eocene.* At Alum Bay the eroded surface of the chalk may be seen with rolled flints lying upon it, and rounded hollows or pot-holes, the appearance being that of a foreshore worn in a horizontal ledge of rock, much like the Horse Ledge at Shanklin.

The land sank again, but not to anything like the depth of the great Chalk Sea. We now come to an era called the Tertiary. The whole geological history is divided

* See Memoir of Geological Survey of I. W. by H. J. Osborne White, F.G.S. 1921, p. 90.

54

into four great eras. The first is the Eozoic, or the age of the Archæan,—often called Pre-Cambrian—rocks; rocks largely volcanic, or greatly altered since their formation, showing only obscure traces of the life which no doubt existed. Then follow the Primary era, or, as it is generally called, the Palæozoic; the Secondary or Mesozoic; and the Tertiary or Kainozoic. Palæozoic is used rather than Primary, as this word is ambiguous, being also used for the crystalline rocks first formed by the solidification of the molten surface of the earth. But Secondary and Tertiary are still in constant use. These long ages, or eras, were of very unequal duration; yet they mark such changes in the life of animal and plant upon the earth that they form natural divisions. The Palæozoic was an immense period during which life abounded in the seas,—numberless species of mollusca, crustaceans, corals, fish are found,—and there were great forests, which have formed the coal measures, on land,— forests of strange primeval vegetation, but in which beautiful ferns, large and small, flourished in great numbers. The Secondary Era may be called the age of reptiles. To this era all the rocks we have so far studied belong. Now we come to the last era, the Tertiary, the age of the mammals. Instead of reptiles on land, in sea and air, we find a complete change. The earth is occupied by the mammalia; the air belongs to the birds such as we see to-day. The strange birds of the Oolitic and Cretaceous have passed away. Birds have taken their modern form. In some parts of the world strata are found transitional between the Secondary and Tertiary.

The Tertiary is divided into four divisions,—the Eocene, the Oligocene (once called Upper Eocene), the Miocene, and the Pliocene; which words signify,—Pliocene the more recent period, Miocene the less recent, Eocene the dawn of the recent.

In the Eocene we shall find marine deposits of a comparatively shallow sea, and beds deposited at the mouth of great rivers, where remains of sea creatures are mingled with those washed down from the land by the rivers. These strata run through the Isle of Wight from east to west, and we may study them at either end of the Island, in Whitecliff and Alum Bays. The strata are highly inclined, so that we can walk across them in a short walk. Some beds contain many fossils, but many of the shells are very brittle and crumbly; and we can only secure good specimens by cutting out a piece of the clay or sand containing them, and transferring them carefully to boxes, to be carried home with equal care. Often much of the face of the cliff is covered with slip or rainwash, and overgrown with vegetation. Sometimes a large slip exposes a good hunting ground.

Now let us walk along the shore, and try to read the story these Tertiary beds tell us. We will begin in Whitecliff Bay. Though easily accessible, it remains still in its natural beauty. The sea washes in on a fine stretch of smooth sand sheltered by the white chalk wall which forms the south arm of the bay. North of the Culver downs the cliffs are much lower, and consist of sands and clays of varying colour, following each other in vertical bands. Looking along the line of shore we notice a band of limestone, at first nearly vertical like the preceding strata, then curving at a sharp angle as it slopes to the shore, and running out to sea in a reef known as Bembridge Ledge. This is the Bembridge limestone; and the beginning of the reef marks the northern boundary of Whitecliff Bay, the shore, however, continuing in nearly the same line to Bembridge Foreland, and showing a continuous succession of Eocene and Oligocene strata. The strata north of the limestone are nearly horizontal, dipping slightly to the north. In the Bembridge limestone we see the end of the Sandown anticline, and the

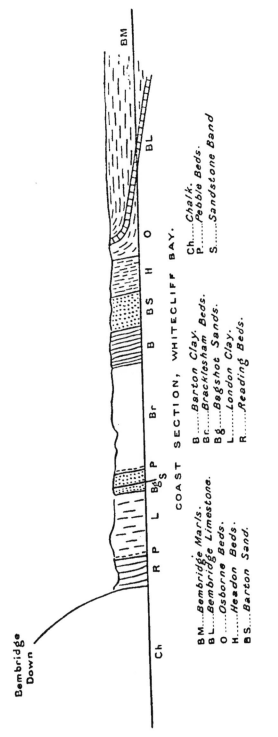

COAST SECTION, WHITECLIFF BAY.

BM....Bembridge Marls.
BL....Bembridge Limestone.
O.......Osborne Beds.
H.......Headon Beds.
BS....Barton Sand.

B ....Barton Clay.
Br.....Bracklesham Beds.
Bg....Bagshot Sands.
L.......London Clay.
R.......Reading Beds.

Ch....Chalk.
P.......Pebble Beds.
S.......Sandstone Band.

FIG. 3

beginning of the succeeding syncline. The strata now dip under the Solent, and rise into another anticline in the Portsdown Hills. North and south of the great anticline of the Weald of Kent and Sussex are two synclinal troughs known as the London and Hampshire basins. Nearly the whole of our English Eocene strata lies in these two basins, having been denuded away from the anticlinal arches. The Oligocene only occur in the Hampshire basin, the higher strata only in the Isle of Wight.

Above the Chalk we come first to a thick red clay called Plastic clay. It is much slipped, and the slip is overgrown. The only fossils found in the Island are fragments of plants ; larger plant remains on the mainland show a temperate climate. This clay was formerly worked at Newport for pottery. The clay is probably a freshwater deposit formed in fairly deep water. On the mainland we find on the border shallow water deposits called the Woolwich and Reading beds. (The clay is 150 to 160 ft. thick at Whitecliff Bay, less than 90 ft. at the Alum Bay.) We come next to a considerable thickness of dark clay with sand, at the surface turned brown by weathering. This is the London clay, so called because it underlies the area on which London is built. At the base is a band of rounded flint pebbles, which extends at the base of the clay from here to Suffolk. In it, as well as in a hard sandstone 18 inches higher up, are tubular shells of a marine worm, *Ditrupa plana*. The sandstone runs out on the shore. About 35 ft. above the basement bed is a zone of *Panopœa intermedia* and *Pholadomya margaritacea*, at 50 ft. another band of *Ditrupa*, and at about 80 ft. a band with a small *Cardita*. In the higher part of the clay are large septaria,—rounded blocks of a calcareous clay-ironstone, with cracks running through them filled with spar. *Pinna affinis* is found in the septaria. The thickness of the clay in Whitecliff Bay is 322 feet.

It can be seen on the shore, when the tide happens to
have swept the sand away. Otherwise the lower beds are
hardly visible, there being no cliff here, but a slope overgrown with vegetation.

In Alum Bay the London clay, about 400 ft. in thickness, consists of clays, chiefly dark blue, with sands, and
lines of septaria. In the lower part is a dark clay with
Pholadomya margaritacea, still preserving the pearly
nacre. There are also *Panopæa intermedia*, and in septaria *Pinna affinis*. All these with their pearly lustre,
are beautiful fossils. A little higher is a zone with
Ditrupa, and further on a band of *Cardita*. Other shells
also are found in the clay, especially in the lower part.
They are all marine, and indicate a sub-tropical climate.
Lines of pebbles show that we are near a beach. In other
parts of the south of England remains from the land are
found, borne down an ancient river in the way we found
before in the Wealden deposits.

But times have changed since the Wealden days, and
the life of the Tertiary times has a much more modern
appearance. From leaves and fruits borne down from
the forest we can learn clearly the nature of the early
Eocene land and climate. Leaves are found at Newhaven,
and numerous fossil fruits at Sheppey. The character of
the vegetation most resembled that now to be seen in India,
South Eastern Asia, and Australia. Palms grew luxuri-
antly, the most abundant fruit being that of one called
Nipadites, from its resemblance to the Nipa palm, which
grows on the banks of rivers in India and the Philippines.
The forests also included plants allied to cypresses,
banksia, maples, poplars, mimosa, custard apples, gourds,
and melons. The rivers abounded in turtle—large
numbers of remains of which are found in the London
clay at the mouth of the Thames—crocodiles and alligators. With the exception of the south east of England,
all the British Isles formed part of a continental mass of

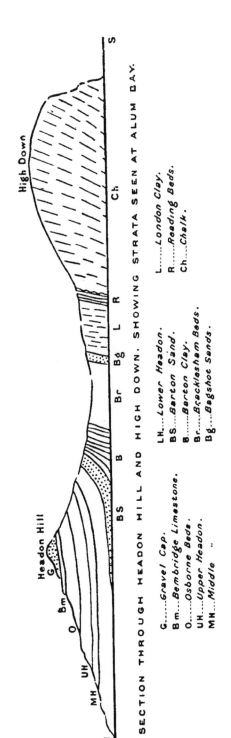

High Down

Headon Hill

G.....Gravel Cap.
Bm...Bembridge Limestone.
O......Osborne Beds.
UH...Upper Headon.
MH...Middle   "

LH...Lower Headon.
BS...Barton Sand.
B......Barton Clay.
Br.....Bracklesham Beds.
Bg....Bagshot Sands.

L......London Clay.
R......Reading Beds.
Ch...Chalk.

SECTION THROUGH HEADON HILL AND HIGH DOWN, SHOWING STRATA SEEN AT ALUM BAY.

FIG. 4

land covered with a tropical vegetation. The mountain chains of England, Scotland, and Wales rose as now, but higher. Long denudation has worn them down since. In the south-east of England the coast line fluctuated ; and sea shells, and the remains of the plant and animal life of the neighbourhood of a great tropical river alternate in the deposits.

The London clay is succeeded by a great thickness of sands and clays which form the Bagshot series. These are divided in the London basin into Lower, Middle, and Upper Bagshot. In the Hampshire basin the strata are now classified as Bagshot Sands, Bracklesham Beds, Barton Beds, the last comprising the Barton Clay and the Barton Sand, formerly termed Headon Hill Sands. There is some uncertainty as to the manner in which these correspond to the beds of the Bagshot district, as the Tertiary strata have been divided by denudation into two groups, and differ in character in the two areas. It is possible that the Barton Sand represents a later deposit than any in the London area.

Almost the only fossil remains in the Bagshot Sands are those of plants, but these are of great interest. In Whitecliff Bay the beds consist for the most part of yellow sands, above which is a band of flint pebbles, which has been taken as the base of the Bracklesham series, for in the clay immediately above marine shells occur. The Bagshot Sands, in Whitecliff Bay, are about 138 feet thick, in Alum Bay, 76 feet, according to the latest classification. In Alum Bay the strata consist of sands, yellow, grey, white, and crimson, with clays, and bands of pipe clay. This is remarkably white and pure, as though derived from white felspar, like the China clay in Cornwall. The pipe clay contains leaves of trees, sometimes beautifully preserved. Specimens are not very easy to obtain, as only the edges of the leaves appear at the surface of the cliff. They have been found chiefly

in a pocket, or thickening of the seam of pipe clay, which
for forty years yielded specimens abundantly, afterwards
thinning out, when the leaves became rare.   The leaves
lie flat, as they drifted and settled down in a pool.   With
them are the twigs of a conifer, occasionally a fruit or
flower, or the wing case of a beetle.   The leaves show a
tropical climate.   The flora is a local one, differing considerably from those of Eocene deposits elsewhere.   The
plants are nearly all dicotyledons.   Of palms there are
only a few fragments, while the London clay of Sheppey
is rich in palm fruits, and many large palms are found
in the Bournemouth leaf beds, corresponding in date to
the Bracklesham.   The differences may be largely due to
conditions of locality and deposition.   The Alum Bay
flora is characterised by a wealth of leguminous plants,
and large leaves of species of fig (*Ficus*) ; simple laurel
and willow-like leaves are common, of which it is difficult
to determine the species, and there is abundance of a
species of *Aralia*.   The character of the flora resembles
most those of Central America and the Malay Archi-
pelago.

The Bracklesham Beds in Alum Bay (570 ft. thick)
consist of clays, with lignite forming bands 6 in. to 2 ft.
thick ; white, yellow, and crimson sands ; and in the
upper part dark sandy clays, with bands showing impressions of marine fossils.   Alum Bay takes its name from
the alum formerly manufactured from the Tertiary clays.
The coloured sands have made the bay famous. The colours
of the sands when freshly exposed, and of the cliffs when wet
with rain, are very rich and beautiful,—deep purple, crimson,
yellow, white, and grey.   Some of the beds are finely
striped in different shades by current bedding.   The
contrast of these coloured cliffs with the White Chalk,
weathered to a soft grey, of the other half of the bay is
very striking and beautiful.   About 45 ft. from the top is
a conglomerate of flint pebbles, some of large size,

NUMMULITES
LÆVIGATUS

TURRITELLA
IMBRICATARIA

LIMNÆA
LONGISCATA.

CARDITA PLANICOSTA

(FUSUS)
LEIOSTAMA PYRUS

CYRENA SEMISTRIATA

PLANORBIS
EUOMPHALUS

EOCENE AND OLIGOCENE

cemented by iron oxide. In Whitecliff Bay the Bracklesham Beds (585 ft.) consist of clays, sands, and sandy
clays, mostly dark, greenish and blue in colour, containing
marine fossils and lignite. Sir Richard Worsley, in his
History of the Isle of Wight, tells that in February, 1773,
a bed of coal was laid bare in Whitecliff Bay, causing
great excitement in the neighbourhood. People flocked
to the shore for coal, but it proved worthless as fuel. It
has, however, been worked to some extent in later years.
In some of the beds are many fossils. Numbers have
lately been visible where a large founder has taken place.
There are large shells of *Cardita planicosta* and *Turritella
imbricataria*. They are, however, very fragile. In a
stratum just above these are numbers of a large Nummulite (*Nummulites lævigatus*). These are round flat shells
like coins,—hence the name (Lat. *nummus*, a coin). They
are a large species of foraminifera. We may split them
with a penknife ; and then we see a pretty spiral of tiny
chambers. A smaller variety, *N. variolarius*, occurs a
little further on, and a tiny kind, *N. elegans*, in the
Barton clay. One of the most striking features of
the later Eocene is the immense development of
Nummulite limestones—vast beds built up of the
delicate chambered shells of Nummulites,—which extend
from the Alps and Carpathians into Thibet, and from
Morocco, Algeria, and Egypt, through Afghanistan and
the Himalaya to China. The pyramids of Egypt are
built of this limestone.

The Bracklesham beds are followed by the Barton clay,
famous for the number of beautiful fossil shells found
at Barton on the Hampshire coast. At Whitecliff Bay
the fossils are, unfortunately, very friable. At Alum
Bay the pathway to the shore is in a gully in the upper
part of the Barton clay. The strata consist of clays,
sands, and sandy clays. The base of the beds is marked
by the zone of *Nummulites elegans*. Numerous very

pretty shells of the smaller Barton types may be found, with fragments of larger ones ; or a whole one may be found. Owing to the cliff section cutting straight across the strata, which are nearly vertical, there is far less of the beds open to observation than at Barton, which probably accounts for the list of fossils being much smaller. The shells are chiefly several species of *Pleurotoma, Rostellaria, Fusus, Voluta, Turritella, Natica*, a small bivalve *Corbula pisum*, a tubular shell of a sand-boring mollusc *Dentalium, Ostræa, Pecten, Cardium, Crassatella*. ¡The fauna is like a blending of Malayan and New Zealand forms of marine life. Throughout the Eocene from the London clay onward the shells are such as abound in the warm sea south east of Asia. Similarly the plant remains take us into a tropic land, where fan palms and feather palms overshadowed the country, trees of the tropics mingling with trees we still find in more Northern latitudes. The general character of the flora as of the shells was Oriental and Malayan ; both being succeeded in later strata by a flora and fauna with greater analogy to that now existing in Western North America.

In Alum Bay the Barton clay is suddenly succeeded by the very fine yellow and white sands which run along the western base of Headon Hill, the curve of the syncline bringing them round from a nearly vertical to an almost horizontal position. These are now known as the Barton Sand. They are 90 ft. thick, the whole of the Barton beds being 338 ft. in Alum Bay, 368 ft. in Whitecliff. The sands were formerly extensively used for glass making. They are almost unfossiliferous. The passage from Barton clay to the sands in Whitecliff Bay is more gradual. The sands here show some fine colouring which reminds us of the more celebrated sands of Alum Bay.

CHAPTER IX

THE OLIGOCENE

WE pass on to strata which used to be called Upper Eocene, but are now generally classified as a period by themselves, and called the Oligocene. They are also known as the Fluvio-marine series. Large part was deposited in fresh water by rivers running into lagoons, or in the brackish water of estuaries, while at times the sea encroached, and beds of marine origin were laid down.

The west of the Island is much the best locality for the lower strata, those which take their name from Headon Hill between Alum and Totland Bays. There are three divisions of the Headon strata, marine beds in the middle coming between upper and lower beds formed in fresh and brackish water. Light green clays are very characteristic of these beds, and at the west of the Island thick freshwater limestones, which have died out before the strata re-appear in Whitecliff Bay. The strongest masses of limestone in Headon Hill belong to the Upper division. The limestones are full of freshwater shells, nearly all the long spiral Limnæa and the flat spiral disc of Planorbis, perhaps the most abundant species being *L. longiscata* and *P. euomphalus*. The limestones themselves are almost entirely the produce of a freshwater plant *Chara*, which precipitates lime on its tissues, in the same manner as the sea weeds we call corallines. On the shore round the base of Headon Hill lie numerous blocks of limestone, the débris of strata fallen in confusion, in which are beautiful specimens of Limnæa and Planorbis. The shells, however, are very fragile. The marine beds of the Middle Headon

63

are best seen in Colwell Bay, where a few yards north
of How Ledge they descend to the beach, and a cliff
is seen formed of a thick bed of oysters, *Ostrea velata*.
The oysters occupy a hollow eroded in a sandy clay
full of *Cytherea incrassata*, from which the bed is known
as the " Venus " bed, the shell formerly being called
Venus, later *Cytherea*, at present *Meretrix*. The marine
beds contain many drifted freshwater shells as Limnæa
and Cyrena. The How Ledge limestone forms the top
of the Lower Headon. It is full of well-preserved
Limnæa and Planorbis.

The Upper and Lower Headon are mainly fresh or
brackish water deposits. The purely freshwater beds
contain *Limnæa, Planorbis, Paludina, Unio*, and land-
shells. In the brackish are found *Potamomya, Cyrena,
Cerithium (Potamides), Melania* and *Melanopsis*. *Paludina lenta* is very abundant throughout the Oligocene.
A large number of the marine shells of the Headon beds
are species also found in the Barton clay. *Cytherea,
Voluta, Ancillaria, Pleurotoma, Natica* are purely marine
genera.

In White Cliff Bay the beds are mostly estuarine. Most
of the fossils are found in two bands, one about 30 ft.
above the base of the series, the other a stiff blue clay,
about 90 feet higher, which seems to correspond with the
" Venus Bed " of Colwell Bay. Many of the fossils are of
Barton types.

The Headon beds are about 150 feet thick at Headon
Hill, 212 ft. in Whitecliff Bay ; and are followed by beds
varying from about 80 to 110 ft. in thickness, known as
the Osborne and St. Helens series. They consist mainly
of marls variously coloured, with sandstone and limestone. In Headon Hill is a thick concretionary limestone, which almost disappears northward. The Oligocene strata often vary considerably within short distances.
The Osborne beds are exposed along the low shore between

Cowes and Ryde, and from Sea View to St. Helens. In Whitecliff Bay they are not well seen, occurring in overgrown slopes. They consist mostly of red and green clays. A band of cream-yellow limestone a foot thick is the most conspicuous feature. The fossils resemble those from the Headon beds, but are much less plentiful. The marls seem to have been mostly deposited in lagoons of brackish water, which at the present day are favourite places for turtles and alligators, and of these many remains are found in the Osborne beds. The beds are specially noted for the shoals of small fish, *Diplomystus vectensis (Clupea)*, first observed by Mr. G. W. Colenutt, F.G.S., and prawns found in them, and also remains of plants. The beds that appear in the neighbourhood of Sea View and St. Helens are divided into Nettlestone Grits and St. Helen's Sands, the former containing a freestone 8 feet thick.

Above these beds lies the Bembridge limestone, which is so conspicuous in Whitecliff Bay, and forms Bembridge Ledge. On the north shore of the Island the strata rise slightly on the northern side of the syncline. There are also minor undulations in an east and west direction. The result is to bring up the Bembridge limestone at various points along the north shore, where it forms conspicuous ledges—Hamstead Ledge at the mouth of the Newtown river, ledges in Thorness Bay, and Gurnard Ledge. In Whitecliff Bay the limestone, about 25 feet thick, forms the conspicuous reef called Bembridge Ledge. The Bembridge limestone consists of two or more bands of limestone with intercalated clays. It is usually whiter than the Headon limestones, and the fossils occur as casts, the shells being sometimes replaced by calc-spar. The limestone has been much used as a building stone for centuries, not only in the Island, but for buildings on the mainland. The most famous quarries were those near Binstead, from which Quarr, the site of the great Abbey, now almost entirely disappeared, derives its name. From these quarries was obtained much of the stone for Winchester Cathedral and many other ancient buildings. In the old walls and buildings of Southampton the stone may be recognised at once by the casts of the Limnææ it contains. The quarries at Qnarr were noted in more ways than one. In later times the remains of early mammalia, —Palæotherium, Anoplotherium, and others—have been found. The quarries are now abandoned and overgrown. The limestone may be seen inland at Brading, where it forms the ridge on which the Church stands.

The limestone is a freshwater formation, and the fossils are mostly freshwater shells, of the same type as the Headon, Limnæa and Planorbis the most common. There are also land shells, especially several species of Helix, the genus which includes the common snail,—*H. globosa*, very large,—and great species of *Bulimus* (*Amphidromus*) and *Achatina* (*B. Ellipticus*, *A. costellata*). These interesting shells were chiefly obtained in the limestone at Sconce near Yarmouth, a locality now inaccessible, being occupied by fortifications. The land shells have an affinity to species now found in Southern North America. The limestone also abounds in the so-called " seeds " of Chara. The reproductive organs,—the " seeds,"—of this curious water-plant, allied to the lower Algæ, are, like the rest of the plant, encased in carbonate of lime, and are very durable. Large numbers are found in the Oligocene strata. Under the microscope they are seen to be beautifully sculptured in various designs, with a delicate spiral running round them. Above the limestone lie the Bembridge marls, varying in thickness in different localities from 70 to 120 feet. North of Whitecliff Bay they stretch on to the Foreland. They are in the main a freshwater formation, but a few feet above the limestone is a marine band with oysters, *Ostrea Vectensis*. It runs out along the shore, where the oysters may be seen covering the surface. The Lower Marls consist chiefly of variously-coloured clays with many shells, chiefly *Cyrena pulchra*, *semistriata*, and *obovata*, *Cerithium mutabile*, and *Melania muricata* (*acuta*) ; and red and green marls, in which are few shells, but fragments of turtle occur. A little above the oyster bed is a band of hard-bluish septarian limestone. Sixty years ago Edward Forbes remarked on the resemblance of this band to the harder insect-bearing limestones of the Purbeck beds. In a limestone exactly resembling this, and similarly situated in the lower part of the marls in Gurnard and Thorness Bays, numerous insects were afterwards found,—beetles, flies, locusts, and dragonflies, and spiders. Leaves of plants, including palms, fig, and cinnamon, have also been found in this bed, showing that the climate was still sub-tropical. The upper Marls consist chiefly of grey clays with abundance of *Melania turritissima* (*Potamaclis*). The chief shells in the marls are *Cyrena*, *Melania*, *Melanopsis* and *Paludina* (*Viviparus*). They are often beautifully preserved ; the species of Cyrena often retain their colour-markings.

Bembridge Foreland is formed by a thick bed of flint gravel resting on the marls, which are seen again in Priory Bay, where in winter they flow over the sea-wall in a semiliquid condition. They lie above the limestone at Gurnard, Thorness, and Hamstead. West of Hamstead Ledge the whole of the beds crop out on the shore, where beautifully preserved fossils may be collected. Large pieces of drift wood occur, also seeds and fruit. Many fragments of turtle plates may be found. Large crystals of selenite (sulphate of lime) occur in the Marls.

Last of the Oligocene in the Isle of Wight are the Hamstead beds. These strata are peculiar to the Isle of Wight. The Bembridge beds also are not found on the mainland, except a small outlier at Creechbarrow Hill in Dorset. The Hamstead beds consist of some 250 feet of marls, in which many interesting fossils have been found. They cover a large area of the northern part of the Island, largely overlaid by gravels, and are only seen on the coast at Hamstead, where they form the greater part of the cliff, which reaches a height of 210 ft., the top being capped by gravel. In winter the clays become semi-liquid, in summer the surface may be largely slip and rainwash, baked hard by the sun. The lower part of the strata may be best seen on the shore. The strata consist of 225 ft. of freshwater, estuarine, and lagoon beds, with *Unio, Cyrena, Cyclas, Paludina, Hydrobia, Melania, Planorbis, Cerithium* (rare), and remains of turtles, crocodiles, and mammals, leaves and seeds of plants ; and above these beds 31 feet of marine beds with *Corbula, Cytherea, Ostrea callifera, Cuma, Voluta, Natica, Cerithium,* and *Melania.*

Except for the convenience of dividing so large a mass of strata, it would not be necessary to divide these from the Bembridge beds, as no break in the character of the life of the period occurs at the junction. The basement bed of the Hamstead strata is known as the Black Band, 2 feet of clay, coloured black with vegetable matter, with *Paludina lenta* very numerous, *Melanopsis carinata, Limnæa, Planorbis,* a small *Cyclas* (*C. Bristovii*), seed vessels, and lumps of lignite. It rests on dark green marls with *Paludina lenta* and *Melanopsis,* and full of roots. This evidently marks an old land surface. About 65 feet higher is the White Band,—a white and green clay full · of shells, mostly broken. There are bands of tabular ironstone containing *Paludina lenta.* Clay ironstone was formerly collected on the shore between Yarmouth and Hamstead and sent to Swansea to be smelted. The strata consist largely of mottled green and red clays, probably deposited in brackish lagoons. These yield few fossils except remains of turtle and crocodile and drifted plants. The blue clays are much more fossiliferous. Among other plants are leaves of palm and water-lily.

The strata gradually become more marine upwards. The marine beds were called by Forbes the Corbula beds, from two small shells, *C. pisum* and *C. vectensis*, of which some of the clays are full. Remains of early mammalia are found in the Hamstead beds, the most frequent being a hog-like animal, of supposed aquatic habits, Hyopotamus, of which there are more than one species.

The fauna and flora of the Oligocene strata show that the climate was still sub-tropical, though somewhat cooling down from the Eocene. Palms grew in what is now the Isle of Wight. Alligators and crocodiles swam in the rivers. Turtle were abundant in river and lagoon. Specially interesting in the Eocene and Oligocene are the mammalian remains. They show us mammals in an early stage before they branched off into the various families as we know them to-day. The Palæotherium was an animal like the tapir, now an inhabitant of the warmer regions of Asia and America. Recent discoveries in Eocene strata in Egypt show stages of development between a tapir-like animal and the elephant with long trunk and tusks. There were in those days hog-like animals intermediate between the hogs and the hippopotami. There were ancestors of the horse with three toes on each foot. There were hornless ancestors of the deer and antelopes. Many of the early mammals showed characters now found in the marsupials, the order to which the Kangaroo and Opossum belong, members of which are found in rocks of the Secondary Era, and are the only representatives of the mammalia in that age. Some of the early Eocene mammalia are either marsupials, or closely related to them. In the Oligocene we find the mammalian life becoming more varied, and branching out into the various groups we know to-day; while the succeeding Miocene Period witnesses the culmination of the mammalia—mammals of every family abounding all over the earth's surface, in a profusion and variety not seen before—or since, outside the tropics.

Chapter X

BEFORE AND AFTER.—THE ICE AGE.

We have read the story written in the rocks of the Isle of Wight. What wonderful changes we have seen in the course of the long history! First we were taken back to the ancient Wealden river, and saw in imagination the great continent through which it flowed, and the strange creatures that lived in the old land. We saw the delta sink beneath the sea, and a great thickness of shallow water deposits laid down, enclosing remains of ammonites and other beautiful forms of life. Then long ages passed away, while in the waters of a deeper sea the great thickness of the chalk was built up, mainly by the accumulation of microscopic shells. In time the sea bed rose, and new land appeared, and another river bore down fruits to be buried with sea shells and remains of turtles and crocodiles in the mud deposited near its mouth to form the London clay. We followed the alternations of sea and land, and the changing life of Eocene and Oligocene times. We have heard of the early mammalia found in the quarries of Quarr, and have learnt from the leaf beds of Alum Bay that at that time the climate of this part of the world was tropical. Indeed, I think everything goes to prove that through the whole of the times we have been studying,— except perhaps the earliest Eocene, that of the Reading beds,—the climate was considerably warmer than it is at the present day. After all these changes do you not want to know what happened next? Well, at this point we come to a gap in the records of the rocks, not only in the Isle of Wight, but also in the British Isles. The

British Isles, or even England and Wales alone, are almost, if not quite unique in the world in that, in their small extent, they contain specimens of nearly every formation from the most ancient times to the present day. In other parts of the world we may find regions many times this area, where we can only study the rocks of some one period. But just at this point in the story comes a period, —a very important one, too,—the Miocene—of which we have no remains in our Islands. We must hear a little of what happened before we come back to the Isle of Wight again in comparatively recent times.

But, first, perhaps, I had better tell,—just in outline,— something of the earlier history of the world, before any of our Isle of Wight rocks were made. For, if I do not, quite a wrong idea may be formed of the world's history. The time of the Wealden river has seemed to us very ancient. We cannot say how many hundreds of thousands, or rather millions of years have passed since that ancient Wealden age. And you may have thought that we had got back then very near the world's birthday, and were looking at some of the oldest rocks on the globe. But no. We are not near the beginning yet. Compared with the vast ages that went before, our Wealden period is almost modern. We cannot tell with any certainty the comparative time ; but we may compare the thickness of strata formed to give us some sort of idea. Now to the first strata in which fossil remains of living things are found we have in all a thickness of strata some 12 times that of all the rocks we have been studying from Wealden to Oligocene, together with the later rocks, Miocene and Pliocene, not found in the Isle of Wight. And before that there is, perhaps, an equal thickness of sedimentary deposits; though the fossils they, no doubt, once contained have been destroyed by changes the rocks have undergone.

Now let me try to give you some idea of the world's

history up to the point where we began in the Isle of Wight. If we could see back through the ages to the furthest past of geological history, we should see our world,—before any of the stratified rocks were laid down in the seas,—before the seas themselves were made,—a hot globe, molten at least at the surface. How do we know this? Because under the rocks of all the world's surface we find there is granite or some similar rock,—a rock which shows by its composition that it has crystallised from a molten condition. Moreover we have seen that the interior of the earth is intensely hot. And yet all along the earth must be radiating off heat into the cold depths of space, and cooling like any other hot body surrounded by space cooler than itself. And this has gone on for untold ages. Far enough back we must come to a time when the earth was red hot,—white hot. In imagination we see it cooling,—the molten mass solidifies into Igneous rock,—the clouds of steam in which the globe is wrapped condense in oceans upon the surface. The bands of crystalline rock that rise above the primeval seas are gradually worn down by rain and rivers and waves, and the first sedimentary deposits laid down in the waters. And in the waters and on the land life appeared for the first time,—we know not how.

A vast thickness of stratified rocks was formed, which are called Archæan (" ancient "). They represent a time, perhaps, as great as all that has followed. These rocks have undergone great changes since their formation. They have been pressed under masses of over-lying strata, and have come into the neighbourhood of the heated interior of the earth ; they have been burnt and baked and compressed and folded, and acted on by heated water and steam, and their whole structure altered by heat and chemical action. Limestones, *e.g.*, have become marble, with a crystalline structure which has obliterated any fossils they may have once contained. Yet it is probable that, like nearly all later limestones, they are of organic origin. These Archæan rocks cover a large extent of country in Canada. We have some of them in our Islands, in the Hebrides, and north-west of Scotland and in Anglesey, and rising from beneath later rocks in the Malvern Hills and Charnwood Forest.*

The Archæan rocks are succeeded by the most ancient fossiliferous rocks, the great series called the Cambrian, because found, and first studied, in Wales. They consist of very hard rocks, and contain large quantities of slate. They are followed by another series called the Ordovician ; and that by another the Silurian. These three great systems of rocks measure in all some 30,000 ft. of strata. They form the hills of Wales and the English Lake District. They contain large masses of volcanic rocks. We can see where were the necks of old volcanoes, and the sheets of lava which flowed from them. The volcanoes are worn down to their bases now ; and the hills of Wales and the Lakes represent the remains of ancient mountain chains, which rose high like the Alps in days of old, long before Alps or Himalayas began to be made. These ancient rocks contain abundant remains of living things, chiefly mollusca, crustaceans, corals, and other marine organisms, showing that the waters of those ages abounded with life.

We must pass on. Next comes a period called the Devonian, or Old Red Sandstone, when the Old Red rocks of Devon and Scotland were laid down. These contain remains of many varieties of very remarkable fish. A long period of coral seas succeeded, when coral reefs flourished over what was to be England ; and their remains formed the Carboniferous Limestone of Derbyshire and the Mendip Hills. A period followed of

* The older division of the Archæan rocks—the Lewisian gneisse—consists entirely of metamorphic and igneous rocks ; a later division—the Torridonian sandstones—is comparatively little altered, but still unfossiliferous.

immense duration, when over pretty well the whole earth
there seem to have been comparatively low lands covered
with a luxuriant and very strange vegetation. The
remains of these ancient forests have formed the coal
measures, which tell of the most widespread and longest
enduring growth of vegetation the world has seen.
Strange as some of the plants were—gigantic horsetails
and club-mosses growing into trees—many were exquis-
itely beautiful. There were no flowering plants, but the
ferns, many of them tree ferns, were of as delicate beauty
as those of the present day. Many of the ferns bore seeds,
and were not reproduced by spores, such as we see on the
fronds of our present ferns. That is a wonderful story
of plant history, which has only been read in recent years.

After the long Carboniferous period came to an end
followed periods in which great formations of red sandstone were made,—the Permian, and the New Red Sandstone or Trias. During much of this time the condition
of the country seems to have resembled that of the
Steppes of Central Asia, or even the great desert of Sahara
—great dry sandy deserts—hills of bare rock with screes
of broken fragments heaped up at their base,—salt
inland lakes, depositing, as the effect of intense evaporation,
the beds of rock salt we find in Cheshire or elsewhere, in
the same manner as is taking place to-day in the Caspian
Sea, in the salt lakes of the northern edge of the Sahara,
and in the Great Salt Lake of Utah.

At the close of the period the land here sank beneath
the sea—again a sea of coral islands like the South Pacific
of to-day. There were many oscillations of level, or
changes of currents ; and bands of clay, when mud from
the land was laid down, alternate with beds of limestone
formed in the clearer coral seas. These strata form a
period known as the Jurassic, from the large development
of the rocks in the Jura mountains. In England the
period includes the Liassic and Oolitic epochs. The

Liassic strata stretch across England from Lyme Regis in Dorset to Whitby in Yorkshire. Most of the strata we are describing run across England from south-west to north-east. After they were laid down a movement of elevation, connected with the movement which raised the Alps in Europe, took place along the lines of the Welsh and Scotch mountains and the chain of Scandinavia, which raised the various strata, and left them dipping to the south-east. Worn down by denudation the edges are now exposed in lines running south-west to north-east, while the strata dip south-east under the edges of the more recent strata. The Lias is noted for its ammonites, and especially for its great marine reptiles, Ichthyosaurus and Plesiosaurus. The Oolitic Epoch follows—a long period during which the fine limestone, the Bath freestone, was made; the limestones of the Cotswolds, beds of clay known as the Oxford and Kimmeridge clays; and again coral reefs left the rock known as coral rag. In the later part of the period were formed the Portland and Purbeck beds, marine and freshwater limestones, which contain also an old land surface, which has left silicified trunks of trees and stems of cycads.

And now following on these came our Wealden strata, the beginning of the Cretaceous period. You see what ages and ages had gone before, and that when Wealden times came, far back as they are, the world's history was comparatively approaching modern times. We must remember that all these formations, of which we have given a rapid sketch, are of great thickness,—thousands of feet of rock,—and represent vast ages of time. See what we have got to from looking at the shells in the sea cliff! We have come to learn something of the world's old history. We have been carried back through ages that pass our imagination to the world's beginning, to the time of the molten globe, before ever it was cool enough to allow life—we know not how—to begin upon its surface. And Astronomy will take us back into an even more distant past, and show us a nebulous mist of vast extent stretching out into space like the nebulæ observed in the heavens to-day, before sun and planets and moons were yet formed. So we are carried into the infinite of time and space, and questions arise beyond the power of human mind to solve.

Now we have, I hope, a better idea of the position the strata we have been specially studying occupy in the geological history, and shall understand the relation the strata we may find elsewhere bear to those in the Isle of Wight and the neighbouring south of England.

After this sketch of what went before our Island story, we must see what followed at the end of the Oligocene period. We said that there are no strata in the British Isles representing the next period, the Miocene. But it was a period of great importance in the world's history. Great stratified deposits were laid down in France and Switzerland and elsewhere, and it was a great age of mountain building. The Alps and the Himalaya, largely composed of Cretaceous and Eocene rocks, were upheaved into great mountain ranges. It is probable that during much of the period the British Isles were dry land, and that great denudation of the land took place. But in the first part of the period at all events this part of the world must have been under water, and strata have been laid down, which have since been denuded away. For our soft Oligocene strata, if exposed to rain and river action during the long Miocene period and the time which followed, would surely have been entirely swept away. The Miocene was succeeded by the Pliocene, when the strata called the Crag, which cover the surface of Norfolk and Suffolk, were formed. They are marine deposits with sea shells, of which a considerable proportion of species still survive.

We have seen that through the ages we have been

studying the climate was mostly warmer than at the present day. The climate of the Eocene was tropical. The Miocene was sub-tropical and becoming cooler. Palms become rarer in the Upper strata. Evergreens, which form three-fourths of the flora in the Lower Miocene, divide the flora with deciduous trees in the Upper. And through the Pliocene the climate, though still warmer than now, was steadily becoming cooler; till in the beginning of the next period, the Pleistocene, it had become considerably colder than that of the present day. And then followed a time which is known as the great Ice Age, or the Glacial Period,—a time which has left its traces all over this country, and, indeed all over Northern Europe and America, and even into southern lands. The cold increased, heavy snowfalls piled up snow on the mountains of Wales, the Lake District, and Scotland; and the snow remained, and did not melt, and more fell and pressed the lower snow into ice, which flowed down the valleys in glaciers, as in Switzerland to-day. Gradually all the vegetation of temperate lands disappeared, till only the dwarf Arctic birch and Arctic willows were to be seen. The sea shells of temperate climates were replaced by northern species. Animals of warm and temperate climates wandered south, and the Arctic fox, and the Norwegian lemming, and the musk ox which now lives in the far north of America took their place; and the mammoth, an extinct elephant fitted by a thick coat of hair and wool for living in cold countries, and a woolly-haired rhinoceros, and other animals of arctic regions occupied the land. When the cold was greatest, the glaciers met and formed an ice-sheet; and Scotland, northern England and the Midlands, Wales, and Ireland were buried in one vast sheet of ice as Greenland is to-day.

How do we know this? To tell how the story has been read would be to tell one of the most interesting stories of geology. Here we can only give the briefest sketch of this wonderful chapter of the world's history. But we must know a little of how the story has been made out. We have already seen that the changes in plant and animal life point to a change from a hot climate, through a temperate, at last to arctic cold. Again, over the greater part of Northern England the rocks of the various geological periods are buried under sheets of tough clay, called boulder clay, for it is studded with boulders large and small, like raisins in a plum pudding. No flowing water forms such a deposit, but it is found to be just like the mass of clay with stones under the great glaciers and ice sheets of arctic regions ; and just such a boulder clay may be seen extending from the lower end of glaciers in Spitzbergen, when the glacier has temporarily retreated in a succession of warm summers. The stones in our boulder clay are polished and scratched in a way glaciers are known to polish and scratch the stones they carry along, and rub against the rocks and other stones. The rock over which the glacier moves is similarly scratched and polished, and just such scratching and polishing is found on the rocks in Wales and the Lake District. Again, we find rocks carried over hill and dale and right across valleys, it may be half across England. We can trace for great distances the lines of fragments of some peculiar rock, as the granite of Shap in Westmorland ; and even rocks from Norway have been carried across the North Sea, and left in East Anglia. This will just give an idea how we know of this strange chapter in the history of our land. For, by this time it was our land—England—much as we know it to-day ; though at times the whole stood higher above sea level, so that the beds of the Channel and the North Sea were dry land. But, apart from variation of level, the geography was in the main as now.

The ice sheet did not come further south than the Thames valley. What was the country like south of this ? Well, you must think of the land just outside the

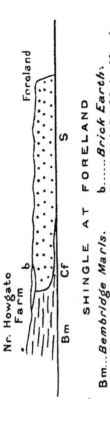

Nr. Howgato Farm

Foreland

Bm

Cf

b

S

SHINGLE AT FORELAND

Bm...Bembridge Marls.　b......Brick Earth.
S......Shingle.　Cf......Old Cliff in Marls.

FIG. 9

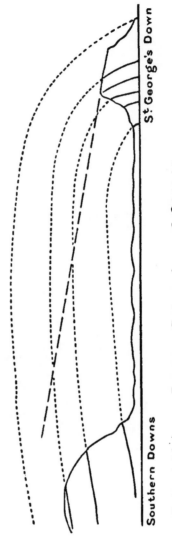

St. George's Down

Southern Downs

FIG. 5

Dotted Lines.....Former Extension of Strata.
Broken Line.....Former Bed of Valley sloping to St. George's Down.

ice sheet in Greenland, or other arctic country. No doubt the winters must have been very severe,—hard frosts and heavy snows,—the ground frozen deep. Some arctic animals would manage to live as they do now just outside the ice sheet in Greenland. Now, have we any deposits formed at that time in the Isle of Wight? I think we have. A large part of the surface of the Island is covered by sheets of flint gravel. The gravels differ in age and mode of formation. We have already considered the angular gravels of the Chalk downs, composed of flints which have accumulated as the chalk which once contained them was dissolved away. But there are other gravel beds, which consist of flints which, after they were set free by the dissolution of the chalk, have been carried down to a lower level by rivers or other agency, and more or less rounded in the process. Many of these beds occur at a high level; and, as they usually cap flat-topped hills, they are known as Plateau Gravels. Perhaps the most remarkable is the immense sheet of gravel which covers the flat top of St. George's Down between Arreton and Newport. Gravel pits show upwards of 30 feet of gravel, consisting of flints with some chert and ironstone, and the greatest thickness is probably considerably more than this. The southern edge of the sheet is cut off straight like a wall. To the north it runs out on ridges between combes which have cut into it. In places in the mass of flints occur beds of sand, which have all the appearance of having been laid down by currents of water. The base of the gravel where it is seen on the steep southern slope of the down has been cemented by water containing iron into a solid conglomerate rock. The flints forming this gravel have not simply sunk down from chalk strata dissolved away; for they lie on the upturned edges of strata from Lower Greensand to Upper Chalk, which have been planed off, and worn into a surface sloping gently to the north; and over this surface

the gravel has somehow flowed.   The sharp wall in which
it ends at the upper part of the slope shows that it once
extended to the south over ground since worn away.
Clearly, the gravel was formed before denudation had cut
out the great gap between the central and southern downs
of the Island.   The down where the gravel lies is 363 ft.
above sea level, 313 ft. above the bottom of the valley
below.   So that, though the gravel sheet is much newer
than the strata we have been studying, it must neverthe-
less be of great antiquity.

It seems that at the top of St. George's Down we are
standing on what was once the floor of an old valley.   In
the course of denudation the bottom of a river valley
often becomes the highest part of a district.   For the bed
of the valley is covered by flint gravel, and flint is ex-
cessively hard, and the bed of flints protects the underlying rock ; so that, while the rocks on each side are worn
away, what was the river bed is eventually left high
above them.   Thus the highest points of a district are
often capped by flint gravel marking the beds of old
streams.   Tracing up this old valley to the southward,
at a few miles distance it will have reached the chalk
region on the south of the anticline : and the flints carried
down the valley may have come from beds of angular
flints already dissolved out of the chalk such as we find
on St. Boniface Down.

But how have these great masses of flints been swept
along ?   Can the land have been down under the sea ;
and have sea waves washed the stones along ?   But these
flints, though water-worn, are not rounded as we find
beach shingle.   What immense rush of water can have
spread these flints 30 feet deep along a river valley ?
We must go to mountain regions for torrents of this
character.   And then, mountain torrents round the stones
in their bed while these are mostly angular.   The history
of these gravels is a difficult one.   I can only give what

seems to me the most probable explanation. It appears to me probable that in the Ice Age, south of the ice sheet, the ground must have been both broken up by frosts, and also held together by being frozen hard to some depth. Then when thaws came in the short but warm summers, or when an intermission of the severe cold took place, great floods would flow down the valleys in the country south of the ice sheet, and masses of ice with frozen earth and stones would be borne along in a sort of semi-liquid flow. In this way Mr. Clement Reid explains the mass of broken-up chalk with large stones found on the heads of cliffs on the South coast, and known by the name of " combe-rock " or " head."

The Ice Age was not one simple period, and it is still difficult to fit together the history we read in different places, and in particular to correlate the gravels of the south of England with the boulder clays of the glaciated area. There were certainly breaks in the period, during which the climate became much milder, or even warm ; and these were long enough for southern species of animals and plants to migrate northward, and occupy the lands where an arctic climate had prevailed. There were moreover considerable variations in the relative level of land and sea. So that we have a very complex history, which is gradually coming into clearer light.

That the gravels of the south of England belong largely to the age of ice, is shown by remains of the mammoth contained in many. These, however, are found in later gravels than those we have considered so far, gravels laid down after the land had been cut down to much lower levels. These lower gravels are known as Valley gravels, because they lie along the course of existing valleys, the Plateau gravels having been laid down before the present valleys came into existence. Teeth of the mammoth are found in the Thames valley, and on the shores of Southampton Water, in gravels about 50 to 70 feet above sea level, and have been found also in the Isle of Wight at Freshwater Gate, at the top of the cliffs near Brook, and in other places. The gravels near Brook with the clays on which they rest have been contorted, and the gravel forced into pockets in the clay, in a manner that suggests the action of grounding ice ploughing into the soil.

The high level gravels must belong to an early stage of the Glacial Epoch. We get some idea of the great length of time this age must have lasted, as we look from St. George's Down over the lower country of the centre of the Island. After the formation of the St. George's Down gravel the vast mass of strata between this and the opposite downs of St. Boniface and St. Catherine's was removed by denudation ; and gravels were then laid down on the lower land, along Blake Down, at Arreton, over Hale common, and along the course of the Yar. Patches of gravel occur on the Sandown and Shanklin cliffs. At Little Stairs a gravel, largely of angular chert, reaches a thickness of 12 feet, and in parts are several feet of loam above gravel.

At the west of the Island a great sheet of gravel covers the top of Headon Hill, reaching a height of 390 feet. It appears sometimes to measure 30 feet in thickness, Like that on St. George's Down it slopes towards the Solent, resting on an eroded surface, in this case of Tertiary strata ; and here too the upper part of the sheet has been removed by the wearing out of the deep valley between the Hill and the Freshwater Downs. The sheet lies on an old valley bottom, which sloped from the chalk downs on the south, then much higher and more extensive than now. Here too we may see something of the length of the Glacial Period. For at Freshwater Gate is a much later gravel, in which teeth of the mammoth have been found. It was probably derived from older gravels that once lay to the south, as the flints are rounded by transport. But the formation of all these gravels appears to

belong to the Glacial Period ; and as we stand in Freshwater Gate, and look at this great gap in the downs worn
out by the Western Yar, and think of the time when a
river valley passed over the tops of the High Downs and
Headon Hill, we receive a strong impression of the length
of the great Ice Age.

Now surely the question will be asked, what caused
these changes of climate in the world's past history—so
that at times a tropical vegetation spread over this land,
and vegetation flourished sufficient to leave beds of coal
within the Arctic circle, and in the Antarctic continent,
and at another the climate of Greenland came down to
England, and an ice sheet covered nearly the whole
country ?   This still remains one of the difficult problems
of Geology.   An explanation has been attempted by
Astronomical Theory, according to which the varying
eccentricity of the earth's orbit—that is to say a slight
change in the elliptic orbit of the Earth, by which at times
it becomes less nearly circular—a change which is known
to take place—may have had the effect of producing these
variations of climatic conditions.   The theory is very
alluring, for if this be the cause, we can calculate mathe-
matically the date and duration of the Glacial Period.
But, unfortunately, supposing the astronomical phenomena to have the effect required, the course of events
given by the astronomical theory would be entirely different to that revealed by geological research.   Geographical explanations have usually failed through being
of too local a character to explain a phenomenon which
affected the whole northern hemisphere, and the effects
of which reached at least as far south as the Equator,*
and are seen again in the southern hemisphere in Australia, New Zealand, and South America.   It is now
believed that great world-movements take place, due to

* The great equatorial mountains Kilimanjaro and Ruwenzori
show signs of a former extension of glaciers.

the contraction by cooling of the Earth's interior, and the adjustment of the crust to the shrinkage.* Possibly some explanation might be found in these world-wide movements ; but their effect seems to last through too long periods of time to suit our Ice Ages. Again, while the geographical distribution of animals and plants in the present and past seems to imply very great changes in the land masses and oceanic areas,† these changes appear to bear no relation to glacial epochs. The cause of the Ice Ages remains at present an unsolved problem. More than one Ice Age has occurred during the long geological history. The marks of such a period are found in Archæan rocks, in the Cambrian, when glaciers flowed down to the sea level in China and South Australia within a few degrees of the tropics, and above all in early Permian times. The Dwyka conglomerate of the Karroo formation of South Africa (deposits of Permo-Carboniferous age) show evidence of extensive glaciation ; deposits of the same age in Northern and Central India, even within the tropics, a glacial series of great thickness in Australia, and deposits in Brazil, appear to show a glaciation greater than that of the recent glacial period. Yet these epochs formed only episodes in the great geological eras. On the whole the climate throughout geological time would seem to have been warmer than at the present day. It may, perhaps, be doubted whether the earth has yet recovered what we may call its *normal* temperature since the Glacial Epoch.

Note on Astronomical Theory.—If the Ice Age be due to the increased eccentricity of the Earth's orbit, the theory shows that a long duration of normal temperature

---

* For an account of such movements, see Prof. Gregory's *Making of the Earth* in the Home University Library.

† See The *Wanderings of Animals*.   By H. Gadow, F.R.S., Cambridge Manuals.

will be followed by a group of Glacial Periods alternating between the northern and southern hemispheres, the time elapsing between the culmination of such a period in one hemisphere and in the other being about 10,500 years. While one hemisphere is in a glacial period, the other will be enjoying a specially mild,—a " genial " period. Now, according to the record of the rocks, the " genial " periods were far from being those breaks in the Glacial which we know as Inter-glacial periods. We have the immensely long warm period of the Eocene and Oligocene, the Miocene with a still warm but reduced temperature, and then the gradual cooling during the Pliocene, till the drop in temperature culminates in the Ice Age. Moreover, the duration of each glaciation during this Ice Age is usually considered to have been much longer than the 10,000 years or so given by the Astronomical Theory. Add to this that the periods of high eccentricity of the Earth's orbit, though occurring at irregular intervals, are, on the scale of geological time, pretty frequent ; so that several of such periods would have occurred during the Eocene alone. Yet the geological evidence shows unbroken sub-tropical conditions in this part of the world throughout the Eocene.

CHAPTER XI

THE STORY OF THE ISLAND RIVERS; AND HOW THE ISLE OF WIGHT BECAME AN ISLAND

WE must now consider the history of the river system of the Isle of Wight, to which our study of the gravels has brought us. For rivers have a history, sometimes a most interesting one, which carries us back far into the past. Even the little rivers of the Isle of Wight may be truly called ancient rivers. For though recent in comparison with the ages of geological time, they are of a vast antiquity compared with the historical periods of human history.

To understand our river systems we must go back to the time when strata formed by deposit of sediment in the sea were upheaved above the sea level. To take the simplest case, that of a single anticlinal axis fading off gradually at each end, we shall have a sort of turtle back of land emerged from the sea, as in figure 6, *aa* being the anticlinal axis. From this ridge streams will run down on either side in the direction of the dip, their course being determined by some minor folds of the strata, or difference of hardness in the surface, or cracks formed during elevation. On each side of the dip-streams smaller ones will flow, more or less in the direction of the strike, and run into the main streams. Various irregularities, such as started the flow of the streams, will favour one or another. Consider three streams, *a, b, c,* and let us suppose the middle one the strongest, with greatest flow of water, and cutting down its bed most rapidly. Its side streams will become steeper and have more erosive force,

FIG. 6

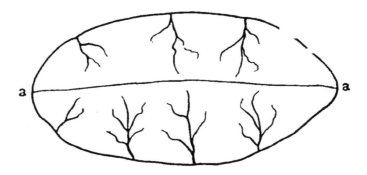

FIG. 7

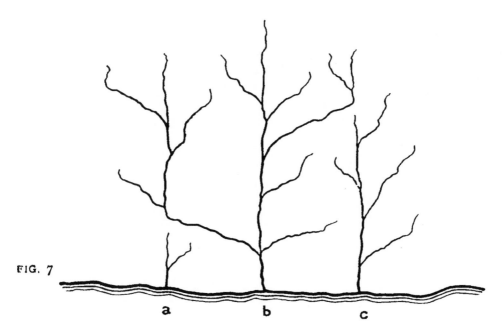

DEVELOPMENT OF RIVER SYSTEMS

and so will eat back their courses most rapidly until they strike the line of the streams on either side.   Their steeper channels will then offer the best way for the upper waters of the streams they have cut to reach the sea ; and these streams will consequently be tapped, and their head waters cut off to flow to the channel of the centre stream. We shall thus have for a second stage in the history a system such as is shown in fig. 7.   The same process will continue till one river has tapped several others ; and there will result the usual figure of a river and its tributaries, to which we are accustomed on our maps. We shall observe that tributaries do not as a rule gradually approach the central stream, but suddenly turn off at nearly a right angle from the direction in which they are flowing, and, after a longer or shorter course, join at another sharp angle a river flowing more or less parallel to their original direction.

The Chalk and overlying Tertiary strata were uplifted from the sea in great folds forming a series of such turtle-backs as we have been considering.   The line of upheaval was not south-west and north-east, as that which raised the older formations in bands across England, but took place in an east and west direction.   The main upheaval was that of the great Wealden anticline.   Other folds produced the Sandown and Brook anticlines, and that of the Portsdown Hills.   The upheaval seemed to have been caused by pressure acting from the south, for the steeper slope of each fold is on the northern side.   Our latest Oligocene strata are tilted with the chalk, showing that the upheaval took place after Oligocene times.   But the great movement was in the main earlier than the Pliocene.   For on the North Downs near Lenham is a patch of Lower Pliocene deposit resting directly on the Chalk, the older Tertiary strata having been removed by denudation, clearly due to the uplift of the Wealden anticline.   The raising of the Pliocene deposit to its

present position proves that the same movement was continued at a later time, probably during the Pleistocene.
But the greater part of the movement may be assigned to
the Miocene, the period of great world-movements which
raised the Alps and the Himalaya.

Many remarkable, and, at first sight, very puzzling
features connected with the courses of rivers find an
explanation when we study the river history. Thus,
looking at the Weald of Kent and Sussex, we see that it
consists of comparatively low ground rising to a line of
heights east and west along the centre, and surrounded
on all sides but the south-east by a wall of Chalk downs.
If we considered the subject, we should suppose that the
drainage of the country would be towards the south-east,
which is open to the sea. Not so. All the rivers flow from
the central heights north and south,—go straight for the
walls of chalk downs, and cut through the escarpment in
deep clefts to flow into the Thames and the Channel.
This is explained when we remember that the rivers began
to flow when the great curve of strata rose above the sea.
Though eroded by the sea during its elevation, yet when
it rose above the waters the arch of chalk must have been
continuous from what are now North Downs to South.
And from the centre line of the great turtle back the
streams began to flow north and south, cutting in the
course of ages deep channels for themselves. The greater
erosion in their higher courses has cut away the mass of
chalk from the centre of the Weald, but the rivers still
flow in the direction determined when the arch was still
entire.

We have a similar state of things in the Isle of Wight.
Any one not knowing the geological story, and looking at
the geography of the Island, might naturally suppose
that there would be a stream flowing from west to east,
through the low ground between the two ranges of downs,
and finding its way into the sea in Sandown Bay. Instead

of this the three rivers of the Island, the two Yars and the Medina, all flow north, and cut through the chalk escarpment of the Central downs, as if an earthquake had made rifts for them to pass, and so find their way into the Solent.   The explanation is the same as in the case of the Weald.   The rivers began to flow when the Chalk strata were continuous over the centre of the Island ; and their course was determined when the east and west anticlinal axis rose above the sea.

We shall notice, however, that the Island rivers start from south of the anticlinal axis.   The centre of the Sandown anticline runs just north of Sandown, but the various branches of the Yar and Medina flow from well south of this.   The explanation would appear to be that the anticline is almost a monoclinal curve,—that is to say, one slope is steep, the other not far from horizontal. Streams starting from the ridge would flow with much greater force down the northern than the southern side, and would cut back their course much more quickly. Thus they would continually cut into the heads of the southern streams, and turn the water supplying them into their own channels.

In its early history a river cuts out its bed, and carries along pebbles, sand and mud to the sea.   The head waters are constantly cutting back, and the slope becoming less steep, till a time comes when the stream in its gently inclined lower course has no more power to excavate, and the finer sediment, which is all that now reaches the lower river, begins to fill up the old channel.   And so the alluvium is formed which fills the lower portions of our river valleys.

Beyond this, the great rush of waters from melting snows and ice of the Glacial Period has come to an end. The gentler and diminished streams of a drier age have no power to roll flint stones along and form beds of gravel. Gravel terraces border our river valleys at a higher level than the present streams. Periods alternated during which gravels were laid down by the river, and when the river acquiring more erosive force, by an elevation of the land giving its bed a steeper gradient, or a wetter climate producing a greater rush of water, cut a new channel deeper in the old valley. So our valleys in Southern England are frequently bordered by a succession of gravel terraces, the higher ones being the older, dating from times when the river flowed at a higher level than at present. Such terraces may be seen above the Eastern Yar and its tributary streams. In the centre of the old gravels is the alluvial flat of a later age.

The Island rivers cut out their channels when the land stood at a higher level than at present. The old channels of the lower parts of the rivers are now filled with alluvium, partly brought down by the rivers and partly marine. The channels are cut down considerably below sea level ; and by the sinking of the land the sea has flowed in, and the last parts of the river courses are now tidal estuaries. The sea does not cut out estuaries. They are the submerged ends of river valleys.

Some idea may be formed of the antiquity of our Island rivers by observing the depth of the clefts they have cut through the downs at Brading, Newport, and Freshwater. But to this we must add the depth at which the old channels lie below the alluvium. It would be interesting to know the thickness of the alluvium. But it is not often that borings come to be made in river alluvia. However, in the old Spithead forts artesian wells are sunk ; and these pass through 70 to 90 feet of recent deposits before entering Eocene strata. Under St. Helen's Fort, at the mouth of Brading Harbour, are 80 feet of recent deposits. The old channel of the Yar, at its mouth, must lie at least at this depth.

Before it passes through the gap in the Chalk downs the Yar has meandered about, and formed the alluvial

flat called Morton marshes. These marshes stretch out
into the flat known as Sandown Level, which occupies the
shore of the bay between Sandown and the Granite Fort.
What is the meaning of this extension of the alluvium
away from the course of the river out to the sea at Sandown? A glance at it as pictured on a geological map
will suggest the answer. We see clearly the alluvia of
two streams converging from right and left, and uniting
to pass to the sea through Brading Harbour. But the
stream to the right has been cut off by the sea encroaching
on Sandown Bay: only the last mile of alluvium is left
to tell of a river passed away. We must reconstruct the
past. We see the Bay covered by land sloping up to east
and south east, the lines of downs extending eastward
from Dunnose and the Culvers, and an old river flowing
northward, and cutting through the chalk at Brading
after being joined by a branch from the west. This old
river must have been the main stream. For it was a
transverse stream, flowing nearly at right angles to the
ridge of the anticline ; while the Yar comes in as a tributary in the direction of the strike. Of other tributary
streams, all from the right are gone by the destruction of
the old land. On the left streams would flow in from the
combes at Shanklin and Luccombe—streams which have
now cut out Shanklin and Luccombe chines.

Passing the gap in the downs the river meandered about,
and, with marine deposit, washed in by the tides, formed
the expanse of alluvium which occupies what was Brading
Harbour,—a harbour which in old times presented at high
tide a beautiful spectacle of land-locked water extending
up to Brading. Inclosures and drainings have been made
from time to time, the upper part near Yarbridge being
taken in in the time of Edward I. Further innings were
made in the reign of Queen Elizabeth ; and Sir Hugh
Middleton, who brought the New River to London,
made an attempt to enclose the whole, but the sea broke

through his embankment. The harbour was finally reclaimed at great cost in 1880, the present embankment enclosing an area of 600 acres.

The history of the Western Yar is similar to that of the Eastern. The main stream must have flowed from land now destroyed by the sea stretching far south of Freshwater Gate. All that is left is its tidal estuary, and the gravel terraces and alluvial flat formed in the last part of its course. Of a tributary stream an interesting relic remains. For more than 2 miles from Chilton Chine through Brook to Compton Grange a bed of river gravel lies at the top of the cliff, marking the course of an old stream, of which coast erosion has made a longitudinal section. This was a tributary of the Yar, when the mammoth left his remains in the gravel at Grange Chine and Freshwater Gate. Down the centre of the gravels lies a strip of alluvium laid down by a stream following the same course in later days. The sea had probably by this time cut into the stream; and it most likely flowed into the sea somewhere west of Brook. In the alluvium hazel nuts and twigs of trees are found at Shippard's Chine near Brook.

The lower course of the Medina is a submerged river valley, the tide flowing up to Newport. The river rises near Chale, and flows through a strip of alluvium, overgrown with marsh vegetation, known as " The Wilderness." This upper course of the Medina, from the absence of gravels or brick earth, has the appearance of a comparatively modern river. But the Medina has a further history. If you look at the map you will see branches of the Yar running south to north as transverse streams, but the main course is that of a lateral river. Look at the two chief sources of the Yar—the stream from near Whitwell and Niton, and that from the Wroxall valley. When they get down to the marshes near Rookley and Merston, they are not flowing at all in the direction of

Sandown or Brading.   They rather look as if they would flow along the marshy flat by Blackwater into the Medina. But the Yar cuts right across their course, and carries them off eastward to Sandown.   When we look, we find a line of river valley with a strip of alluvium running up from the Medina at Blackwater in the direction of these two streams—a valley which the railway up the Yar valley from Sandown makes use of to get to Newport. There can be little doubt that these streams from Niton and Wroxall originally ran along this line into the Medina ; but the Yar, cutting its course backward, has captured them, and diverted their course.   They probably represent the main branches of the Medina in earlier times, the direction of flow from south-east to north-west instead of south to north being possibly due to the overlapping in the neighbourhood of Newport of the ends of the Brook and Sandown anticlines.   The sheet of gravel on Blake Down belongs to this period of the river's history. The river must have diverted between the deposition of the Plateau Gravels and that of the Valley Gravels of the Yar.   For the former follow the original valley, the latter the new course of the river.

We must now take a wider outlook, and see what became of our rivers after they had flowed across what is now the Isle of Wight from south to north.   We have been speaking of times when the Island was of much greater extent than at present.   Standing on the down above the Needles, and looking westward, we see on a clear day the Isle of Purbeck lying opposite, and we can see that the headland there is formed by white chalk cliffs like those beneath us. In front of them stand the Old Harry Rocks, answering to the Needles, both relics of a former extension of the land. In fact Purbeck is just like a continuation of the Isle of Wight.   South of the Chalk lie Greensand and Wealden strata in Swanage Bay, and north towards Poole are Tertiaries.   Clearly these strata were once continuous

with those of the Isle of Wight. We must imagine
the chalk downs of the Island continued as a long range
across what is now sea, and on through Purbeck. A great
Valley must have stretched from west to east, north of
this line, along the course of the Frome, which runs
through Dorset, and now enters the sea at Poole Harbour,
on by Bournemouth, and along the present Solent Channel
—a valley still much above sea level, not yet cut down by
rivers and the sea—and down the centre of this valley a
river must have flowed, which may be ça̧lled the River
Solent. It received as tributaries from the south the
rivers of the Isle of Wight, and others from land since
destroyed by the sea. There flowed into it from the
north the waters of the Stour and Avon, and an old river
which flowed· down the line of what is now Southampton
Water. Southampton Water looks like the valley of a
large river, much larger than the present Test and Itchen.
Its direction points to a river from the north west ; and
it has been shown by Mr. Clement Reid that the Salisbury
rivers—Avon, Nadder, and ₩ily—at a former time, when
they flowed far above their present level—continued their
course into the valley of Southampton Water. For fragments of Purbeck rocks from the Vale of Wardour, west
of Salisbury, have been found by him in gravels on high
land near Bramshaw, carried right over the deep vale of
the Avon in the direction of the Water. The lower Avon
would originally be a tributary of the Solent River ; and
it enters the sea about mid-way between the Needles
and the chalk cliffs of Purbeck, just opposite the point
where we might suppose the sea would have first broken
through the line of chalk downs. No doubt it broke
through a gap made by the course of an old river from the
south, as it is now breaking through the gap made by the
old Yar at Freshwater. When the river Solent had been
tapped at this point, the Avon just opposite would have
acquired a much steeper flow, causing it to cut back at a

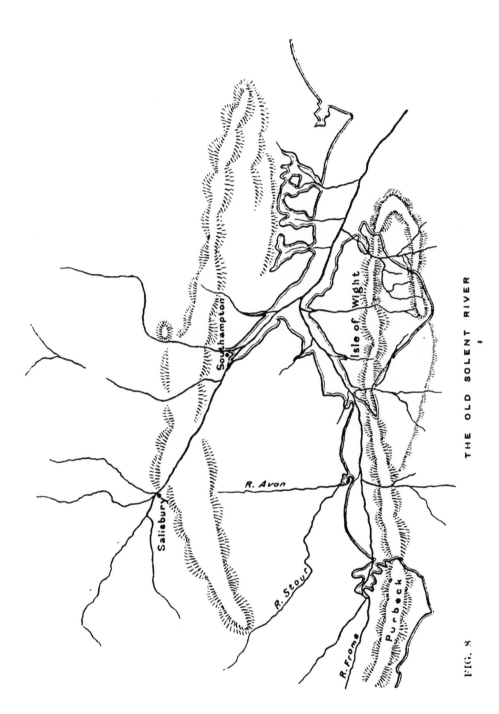

Salisbury

Southampton

Isle of Wight

R. Avon

R. Stour

R. Frome

Purbeck

THE OLD SOLENT RIVER

FIG. 8

faster rate, till it cut the course of the old river which ran by Salisbury to Southampton, and, having a steeper fall, diverted the upper waters of this river into its own channel.

Frost and rain and rivers cut down the valleys of the river system for hundreds of feet ; the sea which had broken through the chalk range gradually cut away the south side of the main river valley from Purbeck to the Needles ; and eventually the valley itself was submerged by a subsidence of the land, and the sea flowed between the Isle of Wight and the mainland.

A gravel of somewhat different character to the rest is the sheet of flint shingle at Bembridge Foreland. It forms a cliff of gravel about 25 feet high resting on Bembridge marls, and consists of large flints, with lines of smaller flints and sand showing current bedding, and also contains Greensand chert and sandstone, which must have been brought from some district beyond the Chalk. The shingle slopes to north-east. To the south-west it ends abruptly, the dividing line between shingle and marls running up steeply into the cliff. This evidently marks an old sea cliff in the marls, against which the gravel has been laid down.*

One or two comparatively recent deposits may be mentioned here. At the top of the cliff in Totland Bay, about 60 ft. above the sea, for a distance of 350 yards, is a lacustrine deposit, consisting in the main of a calcareous tufa deposited by springs flowing from the limestone of Headon Hill. The tufa contains black lines from vegetable matter, and numerous land and freshwater shells of present-day species—many species of Helix, especially *H. nemoralis* and *H. rotundata, Cyclostoma elegans, Limnæa palustris, Pupa, Clausilia, Cyclas,* and others.

On the top of Gore Cliff is a deposit of hard calcareous mud, reaching a thickness of about 9 feet, and forming a small vertical cliff above the slopes of chalk marl. It

*Fig. 9, P. 79.

extends north a few yards beyond the chalk marl on to
Lower Greensand.   It has been formed by rainwash from
a hill of chalk, which must once have existed to the south.
The deposit contains numerous existing land-shells,
especially *Helix nemoralis* and other species of Helix.

Between Atherfield and Chale at the top of the cliff is
a large area of Blown Sand.   The sand is blown up from
the face of the cliff below.   It reaches a thickness of 20
feet, and possibly more in places, and forms a line of
sand dunes along the edge of the cliff.   The upper part of
Ladder Chine shows an interesting example of wind-
erosion.   The sand driven round it by the wind has worn
it into a semi-circular hollow like a Roman theatre.

Small spits, consisting partly of blown sand, extend
opposite the mouths of the Western Yar, the Newtown
river, and the most extensive—at the mouth of the old
Brading Harbour, separating the present reduced Bembridge Harbour from the sea.   This is called St. Helen's
Spit, or " Dover,"—the local name for these sand spits.

CHAPTER XII

THE COMING OF MAN.

WE have watched the long succession of varied life on the earth recorded in the rocks, and now we come to the most momentous event of all in the history—the coming of Man. The first certain evidence of the presence of man on the earth is found with the coming of the Glacial Period,— unless indeed the supposed flint implements found by Mr. Reid Moir, under the Crag in Suffolk, should prove him earlier still. It is a rare chance that the skeleton of a land animal is preserved ; especially rare in the case of a skeleton so frail as that of man. The best chance for the preservation of bones is in deposits in caves, which were frequently the dens of wild beasts and the shelters of man. But the implements used by early man were happily of a very imperishable nature. His favourite material, if he could get it, was flint. Flint could by dexterous blows have flake after flake taken off, till it formed a tool or weapon with sharp point and cutting edge. The implements, though only chipped, or flaked, were often admirably made. They have very characteristic shapes. Moreover, the kind of blow—struck obliquely—by which these early men made their tools left marks which stamp them as of human workmanship. The flake struck off shows what is called a " bulb of percussion "—a swelling which marks the spot where the blow was struck—and from this extends a series of ripples, producing a surface like that of a shell, from which this mode of breaking is called conchoidal fracture. Often, by further chipping the flake itself is worked into an implement. Implements have also been made of chert, but it is far more difficult to work, as it naturally breaks in an irregular way into sharp angular fragments.   Flint, on the other hand, lent itself admirably to the use of early man, who in time acquired a perfect mastery of the material.   The working of flints is so characteristic that, once accustomed to them, you cannot mistake a good specimen.   Sea waves dashing pebbles about will sometimes produce a conchoidal fracture, but never a series of fractures in the methodical way in which a flint was worked by man.   And, of course, specimens may be found so worn that it is difficult to be sure about their nature.   Again early man may, especially in very early times, have been content to use a sharp stone almost as he found it, with only the slightest amount of knocking it into shape. . So that in such a case it will be very difficult to decide whether the stones have formed the implements of man or not.   In later times men learnt to polish their implements, and made polished stone axes like those the New Zealanders and South Sea Islanders used to make in modern times.   The old age of chipped or flaked implements is called the Palæolithic ; the later age when they were ground or polished the Neolithic.   (Simple implements, as knives and scrapers, were still unpolished.) The history of early man is a long story in itself, and of intense interest.   But we must not leave our geological story unfinished by leaving out the culmination of it all in man.   In the higher gravels—the Plateau Gravels— no remains of man are found ; but in the lower—the Valley Gravels,—of the South of England is found abundant evidence of the presence of man.   Large numbers of flint implements have been collected from the Thames valley and over the whole area of the rivers which have gravel terraces along their course.   Over a large sheet of gravel at Southampton, whenever a large gravel pit is dug, implements are found at the base of the gravel.*   The

* Mr. W. Dale, F.S.A.

occurrence of the mammoth and other arctic creatures in the gravels shows that in the Glacial Period man was contemporary with these animals. Remains in caves tell the same story. In limestone caverns in Devon, Derbyshire, and Yorkshire, implements made by man are found in company with remains of the cave bear, cave hyæna, lion, hippopotamus, rhinoceros, and other animals either extinct or no longer inhabitants of this country— remains which have been preserved under floors of stalag-mite deposited in the caves. In caves of central France men have left carvings on bone and ivory, representing the wild animals of that day—carvings which show a remark-able artistic sense, and a keen observation of animal life. Among them is a drawing of the mammoth on a piece of mammoth ivory, showing admirably the appearance of the animal, with his long hair, as he has been found preserved in ice to the present day near the mouths of Siberian rivers. Drawings of the reindeer, true to life, are frequent.

Till recently very few Palæolithic implements had been recorded as found in the Isle of Wight. In the Memoir of the Geological Survey (1889) only one such is recorded, found in a patch of brick earth near Howgate Farm, Bembridge.* A few more implements, which almost certainly came from this brick-earth, have been found on the shore since. In recent years a large number of Palæolithic implements have been found at Priory Bay near St. Helen's. They were first observed on the beach by Prof. E. B. Poulton, F.R.S., in 1886, and were traced to their source in the gravel in the cliff by Miss Moseley in 1902. From that time, and especially from 1904 onwards, many have been found by Prof. Poulton, by R. W. Poulton (and others). Up to 1909 about 150 implements had been found, and there have been more finds since.†

---

* See figure 9, p. 79.

† See account by R. W. Poulton in F. Morey's "Guide to the Natural History of the Isle of Wight."

The most important finds, besides those at Priory Bay, have been those of Mr. S. Hazzledine Warren at Freshwater, especially in trial borings in loam and clay below the surface soil in a depression of the High Downs, south of Headon Hill, at a level of about 360 ft. O.D., in which a number of Palæolithic tools, flakes, and cores were found*. Isolated implements have been found in recent years in various localities in the Island. There are references to finds of implements at different times in the past, but the descriptions are generally too vague to conclude certainly to what date they belong. Much of the gravel used in the Island comes from the angular gravel on St. Boniface Down, or the high Plateau Gravel of St. George's Down ; but in the lower gravels and associated brick earth, it is highly probable that more remains of Palæolithic man will yet be found in the Island, and quite possible that such have been found in the past, but for want of accurate descriptions of the circumstances of the finds are lost to us.

We must pass on to the men of the Neolithic or later stone age. The Palæolithic age was of very great duration, much longer than all succeeding human history. Between Palæolithic and Neolithic times there is in England a large gap. In France various stages have been traced showing a continual advance in culture. In England little, if anything, has been found belonging to the intermediate stages. Such remains may yet be found in caves, or in lower river gravels, now buried below the alluvium. The gap between Palæolithic and Neolithic is marked by the great amount of river erosion which took place in the interval. Palæolithic implements are found in gravels formed when the rivers flowed some 100 feet above their present courses. Take, e.g., the Itchen at Southampton. After the 100 foot gravels were deposited the river cut down, not merely to its present level, but to an old bed now covered up by

* Surv. Mem., I.W., 1921, p. 174.

various deposits beneath the river. After cutting down to that bed the river laid down gravels upon it ; and then—the land standing at a higher level than to-day—the river valley and the sourrounding country were covered by a forest, which, as the climate altered and became damper, was succeeded by the formation of peat. The land has since sunk, and the peat, in parts 17 ft. thick, is now found under Southampton Water, covered by estuarine silt. The Empress Dock at Southampton was dug where a mud bank was exposed at low water. The mud bank was formed of river silt 12 to 17 feet thick. Below this was the peat, resting on gravel. On the gravel horns of reindeer were found. In the peat were large horn cores of the great extinct ox, *Bos primigenius*, also horns of red deer, and also in the peat were found neolithic flint chips, a circular stone hammer head, with a hole bored through for a wooden handle, and a large needle made of horn. Here, at a great interval of time after Palæolithic man, as we see by the history of the river we have just traced, we come to the new race of men, the Neolithic.

When Neolithic man appeared the land stood higher than at present, though not so high as during great part of the Pleistocene. Britain was divided from the Continent, but the shores were a good way out into what is now sea round the coasts, and forests clothed these further shores. Remains of these, known as submerged forest, are found below the tide mark round many parts of our coast. Peat as at Southampton Docks, is found under the estuarine mud off Netley. The wells at the Spithead Forts show an old land surface with peat more than 50 feet below the tide level. The old bed of the Solent river lies much lower still—124 feet below high tide at Noman's Land Fort ; this channel was probably an estuary after the subsidence of the land till it silted up with marine deposits to the level on which the submerged forest grew.

When the Solent and Southampton Water were wooded

H

valleys with rivers flowing down the middle, the Isle of Wight rivers were tributaries to the Solent river, and the forest, as might be expected, extended up their valleys, and covered the low ground of the Island. Under the alluvial flats are remains of buried forests. In digging a well at Sandford in 1906 large trunks of hard oak were found blocking the sinking of the well. When the land sank the sea flowed up the river valleys, converting them into strait and estuary, and largely filling up the channels with the silt, which now covers the peat. In the silt of Newtown river are found bones of *Bos primigenius*, which was found with the Neolithic remains In the peat of Southampton docks.

The remains of Neolithic man are not only found in submerged forests, but over the present surface of the land, or buried in recent deposits. He has left us the tombs of his chiefs, known as long barrows—great mounds of earth covering a row of chambers made of flat stones, such as the mounds of New Grange in Ireland, and the cromlechs or dolmens still standing in Wales and Cornwall. These consist of a large flat or curved stone—it may be 14 feet in length,—supported on three or four others. Originally a great mound of earth or stones was piled on top. These have generally been removed since by the hand of later man. The stones have been taken for road metal, the earth to lay on the land. The great cromlech at Lanyon in Cornwall was uncovered by a farmer, who had removed 100 cart loads of earth to lay on his stony land before he had any idea that it was not a natural mound. Then he came on the great cromlech underneath. Another form of monument was the great standing stone or menhir, one of which, the Longstone on the Down above Mottistone still stands to mark the tomb of some chieftain of, it may be, 4,000 years ago.

The implements of Neolithic man are found all over England, the smooth polished axe head, commonly called a celt (Lat. *celtis*, a chisel), the chipped arrow head, the flaked flint worked by secondary chipping on the edge into a knife, or a scraper for skins; and much more common than the implement, even of the simplest description, are the waste flakes struck off in the making. Very few stone celts have been found in the Isle of Wight. The flakes are extremely numerous, and a scraper or knife may often be found. They are turned up by the plough on the surface of the fields, in the earth of which they have been preserved from rubbing and weathering. They have however, acquired a remarkable polish, or " patina "—how is not clearly explained—which distinguishes their surface from the waxy appearance of newly-broken flint. In places the ground is so covered with flakes that we can have no doubt that these are the sites of settlements. The implements were made from the black flints fresh out of the chalk, and we can locate the Neolithic flint workings. In our northern range of downs where the strata are vertical the layers of flint in the Upper Chalk run out on the top of the downs, only covered with a thin surface soil. In places where this soil has been removed—as in digging a quarry— the chalk is seen to be covered with flakes similar to those found on the lower ground, save that they are weathered white from lying exposed on the hard chalk, instead of on soft soil into which they would gradually sink by the burrowing of worms. It is probable that these flakes would be found more or less along the range of downs under the surface soil.

In places on the Undercliff have been found what are known as Kitchen Middens—heaps of shells which have accumulated near the huts of tribes of coast dwellers, who lived on shellfish. One such was formerly exposed in the stream below the old church at Bonchurch, and is believed to extend below the foundations of the Church.

After a long duration of neolithic times a great step in civilisation took place with the introduction of bronze.

Bronze implements were introduced into this country probably some time about B.C. 1800–1500; and bronze continued to be the best material of manufacture till the introduction of iron some two or three centuries before the visit of Julius Cæsar to these Islands.   To the early bronze age belong the graves of ancient chieftains known as round barrows, of which many are to be seen on the Island downs.   Funeral urns and other remains have been found in these, some of which are now in the museum at Carisbrooke Castle.   Belonging to later times are the remains of the Roman villa at Brading and smaller remains of villas in other places;   and cemeteries of Anglo-Saxon date, rich in weapons and ornaments, which have been excavated on Chessil and Bowcombe Downs.   But the study of the remains of ancient man forms a science in itself—Archæology.   In studying the periods of Palæolithic and Neolithic man we have stood on the borderland where Geology and Archæology meet.   We have seen that vast geological changes have taken place since man appeared on earth.   We must remember that the geological record is still in process of being written.   It is not the record of a time sundered from the present day, but continuous with our own times;   and it is by the study of processes still in operation that we are able to read the story of the past.

Chapter XIII.

THE SCENERY OF THE ISLAND—Conclusion.

After studying the various geological formations that enter into the composition of the Isle of Wight, and learning how the Island was made, it will be interesting to take a general view of the scenery, and see how its varied character is due to the nature of its geology. It would hardly be possible to find anywhere an area so small as this little Island with such a variety of geological formations. The result is a remarkable variety in the scenery.

The main feature of the Island is the range of chalk downs running east and west, and terminating in the bold cliffs of white chalk at Freshwater and the Culvers. Here we have vertical cliffs of great height, their white softened to grey by weathering and the soft haze through which they are often seen. In striking contrast of colour are the Red Cliff of Lower Greensand adjoining the Culvers, and the many-coloured sands of Alum Bay joining on to the chalk of Freshwater. The summits of the chalk downs have a characteristic softly rounded form, and the chalk is covered with close short herbage suited to the sheep which frequently dot the green surface. Where sheets of flint gravel cap the downs, as on St. Boniface, they are covered by furze and heather, producing a charming variation from the smooth turf where the surface is chalk. The Lower Greensand forms most of the undulating country between the two ranges of downs; while the Upper Greensand, though occupying a smaller area, produces one of the most conspicuous features of the scenery— the walls of escarpment that form the inland cliffs between

Shanklin and Wroxall, Gat Cliff above Appuldurcombe, the
fine wall of Gore Cliff above Rocken End, and the line of
cliffs above the Undercliff.   To the Gault Clay is due the
formation of the Undercliff—the terrace of tumbled strata
running for miles well above the sea, but sheltered by an
upper cliff on the north, and in parts overgrown with
picturesque woods.   The impervious Gault clay throws
out springs around the downs, which form the headwaters
of the various Island streams. . The upper division of the
Lower Greensand, the Sandrock, forms picturesque undulating foothills, often wooded, as at Apsecastle, and at
Appuldurcombe and Godshill Park.   On a spur of the
Sandrock stands Godshill Church, a landmark visible
for miles around.   At Atherfield we have a fine line of
cliffs of Lower Greensand, while the Wealden Strata on to
Brook form lower and softer cliffs.

To the north of the central downs the Tertiary sands
and clays, often covered by Plateau gravel, form an
extended slope towards the Solent shore, much of it well
wooded, and presenting a charming landscape seen from
the tops of the downs.   This slope of Tertiary strata is
deeply cut into by streams, which form ravines and
picturesque creeks, as Wootton Creek, 200 feet below the
level of the surounding country.   While much of the
Island coast is a line of vertical cliff, the northern shores
are of gentler aspect, wooded slopes reaching to the
water's edge, or meadow land sloping gradually to the
sea level.   Opposite the mouths of streams are banks of
shingle and sand dunes, forming the spits locally known as
" dovers."   Some of these, in particular, St. Helen's Spit,
afford interesting hunting grounds for the botanist.

The great variety of soil and situation renders the Isle
of Wight a place of interest to the botanist.   We have
the plants of chalk downs, of the sea cliff and shore, of the
woods and meadows, of lane and hedgerow, and of the
marshes.   The old villages of the Island, often occupying

very picturesque situations—as Godshill on a spur of the
southern downs, Newchurch on a bluff overlooking the
Yar valley, Shorwell nestling among trees in a south-
looking hollow of the downs, Brighstone with its old church
cottages and farmhouses among trees and meadows
between down and sea—the old and interesting churches,
the thatched cottages, the old manor houses of Elizabethan
or Jacobean date, now mostly farm houses, for which the
Island is famous, add to the varied natural beauty.

One of the most characteristic features of the southern
coasts of the Island, should be mentioned, the Chines,—
narrow ravines which cut inland from the coast through
the sandstone and clays of the Greensand and Wealden
strata, and along the beds of which small streams flow to
the sea. Narrow and steep-sided,—the name by which
they are called is akin to *chink*—they are in striking
contrast to the more open valleys of the streams which
flow into the Solent on the north shore of the Island. The
most beautiful is Shanklin Chine. The cliff at the mouth
of the chine, just inside which stands a picturesque fisher-
man's cottage with thatched roof, is 100 ft. high ; and the
chasm runs inland for 350 yds., to where a very reduced
cascade (for the water thrown out of the Upper Greensand
by the Gault clay is tapped at its source for the town
supply) falls vertically over a ledge produced by hard
ferruginous beds of the Greensand. Above the cascade
the ravine runs on, but much shallower, for some 900 yards.
The lower ravine has much beauty, tall trees rising up the
sides, and overshadowing the chasm, the banks thickly
clothed with large ferns and other verdure. Much wilder
are the chines on the south-west of the Island. The
cascade at Blackgang falls over hard ferruginous beds (to
which the beds over which Shanklin cascade falls—though
on a smaller scale—probably correspond). The chine
above these beds, being hollowed out in the soft clays and
sands of the Sandrock series, is much more open. Whale

Chine is a long winding ravine between steep walls, the stream at the bottom making its way through blocks of fallen strata.

The cause of these chines seems to be the same in all cases. It may be noticed that Shanklin and Luccombe chines are cut in the floors of open combes,—wide valleys with gently sloping floors ; and at each side of these chines is to be seen the gravel spread over the floor of the old valley. It can scarcely be doubted that these combes are the heads of the valleys of the old streams, which flowed down a gradual slope till they joined the old branch (or, rather the old main river)* of the Yar, flowing over land extending far over what is now Sandown Bay. When the sea encroached, and cut into the course of this old river, and on till it made a section of what had been the left slope of the valley, the old tributaries of the Yar now fell over a line of cliff into the sea. They thus gained new erosive power, and cut back at a much greater rate new and deeper channels ; with the result that narrow trenches were cut in the floors of the old gently sloping valleys. The chines on the S.W. coast are to be explained in a similar way. They have been cut back with vertical sides, because the encroachment of the sea caused the streams to flow over cliffs, and so gave then power to cut back ravines at so fast a rate that the weathering down of the sides could not keep pace with it. The remarkable wind-erosion of these bare south-westerly cliffs by a sort of sand-blast driven before the gales to which that stretch of coast is exposed has already been referred to.

A few words in conclusion to the reader. I have tried to show you something of the interest and wonder of the story written in the rocks. We have seen something of the world's making, and of the many and varied forms of life which have succeeded each other on its surface. We have had a glimpse of great and deep problems suggested, which

* See p. 91.

are gradually receiving an answer. Geology has the advantage that it can be studied by all who take walks in the country, and especially by those who visit any part of the sea coast, without the need of elaborate and costly scientific instruments and apparatus. Any country walk will suggest problems for solution. I have tried to lead you to observe nature accurately, to think for yourselves, to draw your own conclusions. I have shown you how to try to solve the questions of geology by looking around you at what is taking place to-day, and by applying this knowledge to explain the records which have reached us of what has happened in the past. You are not asked to accept the facts of the geological story on the word of the writer, or on the authority of others, but to think for yourselves, to learn to weigh evidence, to seek only to find out the truth; whether it be geology you are studying or any other subject, and to follow the truth whithersoever it leads.

TABLE OF STRATA

Recent.   Peat and River Alluvium.

Pleistocene.   Plateau Gravels :   Valley Gravels and Brick-Earth.

Pliocene ⎫
Miocene ⎭   Absent from the Isle of Wight.

Tertiary

Oligocene

Hamstead ⎰ Marine, Corbula **Beds** / Freshwater & Estuarine.

Bembridge Beds ⎰ **Bembridge** Marls / Bembridge Limestone

Osborne and St. Helen's Beds.

Headon Beds ⎰ Upper. Freshwater and Brackish / Middle. Marine / Lower. Freshwater and Brackish

Eocene

Barton Beds ⎰ Barton Sand. / Barton Clay.

Bracklesham Beds.
Bagshot Sands
London Clay
Plastic Clay (Reading Beds)

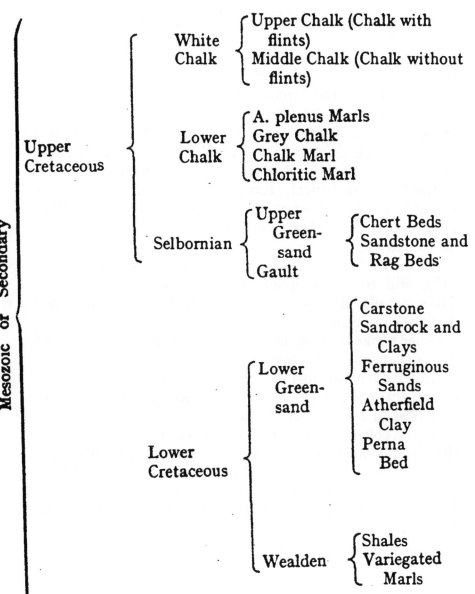

Mesozoic or Secondary

Upper Cretaceous

White Chalk — Upper Chalk (Chalk with flints) / Middle Chalk (Chalk without flints)

Lower Chalk — A. plenus Marls / Grey Chalk / Chalk Marl / Chloritic Marl

Selbornian — Upper Greensand (Chert Beds / Sandstone and Rag Beds) / Gault

Lower Cretaceous

Lower Greensand — Carstone / Sandrock and Clays / Ferruginous Sands / Atherfield Clay / Perna Bed

Wealden — Shales / Variegated Marls

FOR FURTHER STUDY.

Memoirs of the Geological Survey. General Memoir of the Isle of Wight, date 1889. New edition, entitled " A short account of the Geology of the Isle of Wight," by H. J. Osborne White, F.G.S., 1921, price 10s. The Memoirs are the great authorlty for the Geology of the Island : technical ; books for Geologists. The New Edition is more condensed than the original, but contains much later research. Mantell's " Geological Excursions round the Isle of Wight," 1847. By one of the great early geologists. Long out of print, but worth getting, if it can be picked up second-hand.

Norman's " Guide to the Geology of the Isle of Wight," 1887, still to be obtained of Booksellers in the Island. Gives details of strata, and lists of fossils, with pencil drawings of fossils.

Other books bearing on the subject have been mentioned in the text and foot-notes.

An excellent geological map of the Island, printed in colour, scale 1in. to the mile, full of geological information, is published by the Survey at 3s.

A good collection of fossils and specimens of rocks from the various strata of the Isle of Wight has recently been arranged at the Sandown Free Library, and should be visited by all interested in the Geology of the Island. It should prove a most valuable aid to all who take up the study, and a great assistance in identifying any specimens they may themselves find.

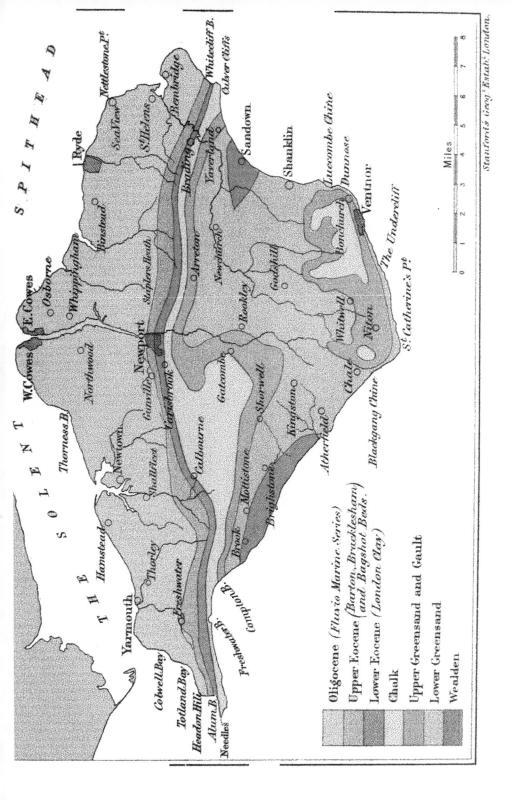

THE SOLENT

SPITHEAD

W. Cowes
E. Cowes
Osborne
Thorness B.
Newtown
Shalfleet
Hamstead
Thorley
Yarmouth
Freshwater
Colwell Bay
Totland Bay
Headon Hill
Alum B.
Needles
Freshwater B. (Compton B.)
Brook
Mottistone
Brighstone
Kingston
Shorwell
Gatcombe
Calbourne
Atherfield
Blackgang Chine
Chale
Kingston
Whitwell
Niton
Rookley
Godshill
St. Catherine's Pt.
The Undercliff
Bonchurch
Ventnor
Luccombe Chine
Dunnose
Shanklin
Sandown
Newchurch
Arreton
Staplers Heath
Newport
Gurnard
Carisbrook
Northwood
Whippingham
Binstead
Ryde
Sea View
Nettlestone Pt.
St. Helens
Bembridge
Whitecliff B.
Culver Cliffs
Yaverland
Brading

Oligocene (Fluvio Marine Series)
Upper Eocene (Barton, Bracklesham)
(and Bagshot Beds.)
Lower Eocene (London Clay)
Chalk
Upper Greensand and Gault
Lower Greensand
Wealden

Miles
0   1   2   3   4   5   6   7   8

Stanford's Geog.¹ Estab.¹ London.

INDEX

Words in Italics refer to a page where the meaning of a term is given.

BV - #0020 - 260522 - C0 - 229/152/9 - PB - 9781333469481 - Gloss Lamination